Six Arguments for a Greener Diet

Six Arguments for a Greener Diet

How a More Plant-Based Diet Could Save Your Health and the Environment

Center for Science in the Public Interest

First Printing, July 2006

2 4 6 8 10 9 7 5 3 1

The Center for Science in the Public Interest (CSPI), founded in 1971, is a nonprofit organization that conducts innovative education, research, and advocacy programs in the area of nutrition, food safety, environment, and alcoholic beverages. CSPI is supported by the 900,000 subscribers in the United States and Canada to its *Nutrition Action Healthletter* and by foundation grants.

Center for Science in the Public Interest
1875 Connecticut Avenue, NW, #300
Washington, DC 20009
Tel: 202-332-9110; fax: 202-265-4954
Email: cspi@cspinet.org; Internet: www.cspinet.org

ISBN 0-89329-049-1

Visit CSPI's Eating Green web site: www.EatingGreen.org.

The Web of Animal-Based Foods and Problems

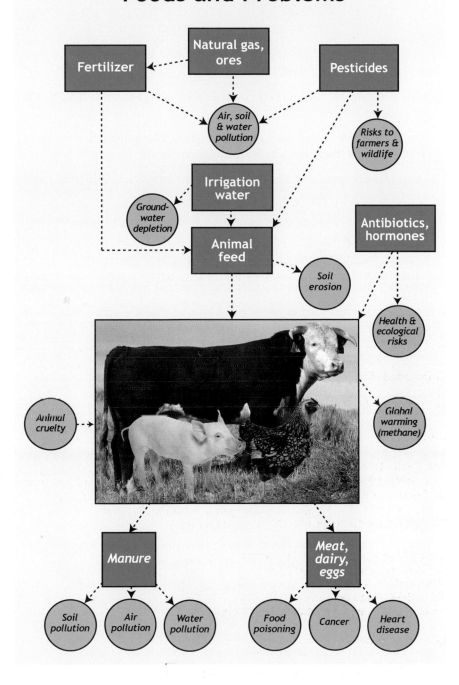

Eating Green: By the Numbers

(All figures apply to the United States, except where noted, and are approximate. See text for sources.)

Health

3 years: how much longer vegetarian Seventh-day Adventists live than non-vegetarian Seventh-day Adventists

4.9 servings: the servings of fruits and vegetables consumed daily, compared to the recommended 5 to 10

16 percent: the decreased mortality from heart disease associated with eating one additional serving of fruits or vegetables each day

24 percent: how much lower the rate of fatal heart attacks is in vegetarians compared to non-vegetarians

25 percent: the proportion of food-poisoning deaths due to pathogens from animals or their manure

33 percent: the decrease in beef consumption since 1976

50 percent: how much less dietary fiber Americans consume than is recommended

51 percent: the reduction in risk of heart attack for people eating nuts five or more times per week compared to less than once a week

90 percent: the proportion of chickens contaminated with *Campylobacter* bacteria

100 percent: how much fattier meat is from a typical steer that's fed grain rather than grass

199 pounds: the combined amount of meat, poultry, and seafood produced per American (2003)

1,100: the mortalities due each year to foodborne illnesses linked to meat, poultry, dairy, and egg products

46,000: the number of illnesses due annually to antibiotic-resistant strains of *Salmonella* and *Campylobacter*

63,000: the number of deaths from coronary heart disease caused annually by the fat and cholesterol in meat, dairy, poultry, and eggs

$7 billion: the annual medical and related costs of foodborne illnesses

$37 billion: the annual cost of drugs to treat high blood pressure, heart disease, and diabetes

$50 billion: the annual cost of coronary bypass operations and angioplasties

Environment

1 pound: the amount of fertilizer needed to produce 3 pounds of cooked beef

5 times as much: the irrigation water used to grow feed grains compared to fruits and vegetables

5 tons: the soil lost annually to erosion on an average acre of cropland

7 pounds: the amount of corn needed to add 1 pound of weight to feedlot cattle (some of that weight gain is not edible meat)

19 percent: the proportion of all methane, a greenhouse gas, emitted by cattle and other livestock

41 percent: the share of irrigated land planted in livestock feed crops

66 percent: the proportion of grain that ends up as livestock feed at home or abroad

331: the number of odor-causing chemicals in hog manure

4,500 gallons: the rain and irrigation water needed to produce a quarter-pound of raw beef

8,500 square miles: the size of the "dead zone" created in the Gulf of Mexico by fertilizer runoff carried by the Mississippi from the upper Midwest

33 million: the number of cars needed to produce the same level of global warming as is caused by the methane gas emitted by livestock and their manure

22 billion pounds: the amount of fertilizer used annually to grow feed grains for American livestock

3.3 trillion pounds: the amount of livestock manure produced annually

17 trillion gallons: the amount of irrigation water used annually to produce feed for U.S. livestock

Animal Welfare

0.5 square feet: the amount of space allotted to the average layer hen

30: the number of chickens and turkeys consumed annually by the average American

13,200: the number of chickens killed each hour in a modern slaughterhouse

50,000: the number of broiler chickens in the largest growing sheds

140 million: the number of cattle, pigs, and sheep slaughtered each year

Contents

Appendixes and Notes

Acknowledgments

This book is a publication of the Center for Science in the Public Interest's (CSPI's) Eating Green project, which advocates a more plant-based diet to protect both health and the environment. Asher Wolf drafted the chapters on foodborne illness, soil, water, air, and animal welfare; Reed Mangels wrote the chapter on chronic disease; and Michael F. Jacobson wrote several other chapters and edited the entire manuscript. Michael Kisielewski contributed valuable editing and research; Moira Donahue, Judy Jacobs, Phyllis Machta, Tyler Martz, Jonathan Morgan, and Carol Touhey helped with proofreading and fact checking. Nita Congress provided invaluable advice while she edited and designed the book. CSPI's Debra Brink designed the cover and several graphic displays.

Numerous experts in government, academe, and nonprofit organizations generously provided data, advice, and reviews of entire chapters. Those people include Tamar Barlam, Aaron Blair, Navis Bermudez, Lawrence Cahoon, Winston Craig, Karen Florini, Tom Gegax, Noel Gollehon, Michael Greger, Robert Hadad, Ed Hopkins, Dennis Keeney, Ronald Lacewell, Alice Lichtenstein, Robbin Marks, Roy Moore, Mark Muller, Frensch Niegermeier, David Pimentel, Nancy Rabalais, Darryl Ray, Steven Roach, Bernard Rollin, Gail Rose, Joe Rudek, Daniel Rule, Frank Sacks, Jennifer Sass, Paul Shapiro, Parke Wilde, and George Wuerthner. In addition, CSPI staffers Caroline Smith DeWaal, Bonnie Liebman, and David Schardt reviewed chapters and offered much useful advice. Notwithstanding all that assistance, this book might still contain factual errors and inappropriate characterizations, for which the editor, Michael F. Jacobson, deserves the dubious credit. Finally, we are grateful to John Robbins for writing his ground-breaking *Diet for a New America*, which helped inspire our work.

CSPI extends its sincere gratitude to the Freed Foundation, Tom and Mary Gegax, the Shared Earth Foundation, Lucy Waletzky, and the Wallace Genetic Foundation for their generous support of the Eating Green project and the preparation of this book.

Abbreviations

AMR	advanced meat recovery
BMI	body mass index
BSE	bovine spongiform encephalopathy
CAFO	concentrated animal feeding operation
CDC	Centers for Disease Control and Prevention
CHIP	Coronary Health Improvement Project
CLA	conjugated linoleic acid
CRP	Conservation Reserve Program
CSPI	Center for Science in the Public Interest
DASH	Dietary Approaches to Stop Hypertension
DHA	docosahexaenoic acid
EDC	endocrine-disrupting compound
EPA	eicosapentaenoic acid
EPA	Environmental Protection Agency
EPIC	European Prospective Investigation into Cancer and Nutrition
EQIP	Environmental Quality Incentives Program
EWG	Environmental Working Group
FDA	Food and Drug Administration
HCA	heterocyclic amine
HDL	high-density lipoprotein

LDL	low-density lipoprotein
PAH	polycyclic aromatic hydrocarbon
PBDE	polybrominated diphenyl ether
PCB	polychlorinated biphenyl
PETA	People for the Ethical Treatment of Animals
ppm	parts per million
rBST	recombinant bovine somatotropin
SDA	Seventh-day Adventist
USDA	U.S. Department of Agriculture
USGS	U.S. Geological Survey
vCJD	variant Creutzfeldt-Jakob disease
VOC	volatile organic compound
WIC	Women, Infants, and Children

Preface:
Greener Diets for a Healthier World

Americans eat what might be called an all-consuming diet. Together, we represent over 40 billion pounds of protoplasm that each day needs to be fed over 1 billion pounds and 1 trillion calories of food. Our agricultural system consumes enormous quantities of fuel, fertilizers, and pesticides to produce the grains, meat and poultry, and fruits and vegetables that feed a nation of 300 million people. It consumes enormous tracts of land and quantities of water—not only for growing food for people, but also for producing food for livestock. And ultimately it consumes the consumer: Diet-related diseases account for hundreds of thousands of premature deaths each year.

Six Arguments for a Greener Diet analyzes the multitudinous and far-reaching effects of livestock production and consumption. On the health front, most consumers probably know that the saturated fat and cholesterol in fatty beef and dairy products and eggs promote heart disease. Fewer people are aware that beef has been linked to colon cancer and milk to prostate cancer. Adding to the toll are the toxic chemicals, such as polychlorinated biphenyls (PCBs), that animals tend to accumulate in their muscle fat and milk. In all, animal foods may be responsible for 50,000 to 100,000 premature deaths each year. (Not surprisingly, vegetarians tend to be healthier than the rest of us.)

While heart disease and cancer generally take decades to develop, meat, poultry, and eggs are major causes of food poisoning, which shows up quickly. Over 1,000 people die each year from livestock-related food-borne illnesses caused by bacteria such as *Salmonella* and *E. coli*. In fact, many foodborne illnesses traced to fruits and vegetables actually are due to animal manure that gets onto crops. Some foodborne germs are especially harmful because they defy the usual antibiotic treatment. Such antibiotic resistance results, in part, from the feeding of small amounts of antibiotics to cattle, hogs, and poultry to fatten the animals faster or compensate for the dirty, crowded conditions in which they live.

Consuming large quantities of animal products has inevitable environmental consequences. Beef cattle typically live out their last several months in huge, densely populated feedlots. The 50,000 cattle that reside in a large feedlot at a given time produce as much manure as a city of several million people. Not surprisingly, they create a stench that undermines the quality of life for everyone who lives or works nearby. Even grazing can be problematic. In some parts of the West, cattle graze on ecologically sensitive land, which can destroy normal vegetation. Industrial-scale hog production relies on pond-sized cesspools (euphemistically called lagoons by agribusiness) of manure. Stench aside, cesspools sometimes break open and pollute local streams and rivers.

A high percentage of the grains and hay grown on our nation's farms feeds animals, not humans. Producing the vast quantities of corn, soybean meal, alfalfa, and other ingredients of livestock feed consumes vast quantities of natural resources and requires thousands of square miles of land. Much of the Midwest's grasslands and forests have been replaced by grain farms. In the arid West and Great Plains, large amounts of irrigation water, which might otherwise be used as drinking water or in more productive commercial enterprises, are needed to produce feed grains. Although shifting to grass-fed beef would solve some of the environmental problems, as well as provide leaner meat, one serious problem would remain: Cattle naturally emit methane, a potent greenhouse gas.

The chemical fertilizers that farmers use to help maximize grain production take a great deal of energy to produce, and they pollute waterways and drinking water. Because of all the fertilizer that washes down the Mississippi River, the Gulf of Mexico has a poorly oxygenated "dead zone" the size of New Jersey. Using chemical pesticides to protect crops from insects and other pests frequently results in those chemicals contaminating drinking water in rural areas, as well as endangering farmworkers and wildlife. The small amounts that we consume when we eat both plant and animal foods are unwelcome, if not demonstrably harmful.

Among the questions this book seeks to answer are "What is the cost to the environment of raising so many food animals?" and "What is the cost to our bodies of eating them?" We also ask "What is the cost to the animals?" If an animal is treated well, can exhibit its natural behaviors, and has a quick and painless death, then killing and eating it is easier to justify. However, most food animals are not so lucky. Hogs' tails and chickens' beaks are partially cut off. Egg-laying hens are squeezed into small cages. Broiler chickens spend their entire short lives in sheds crammed with tens of thousands of birds, never getting a glimpse of the outdoors or pecking for insects in the ground. Steers are often branded with hot irons, and bulls are castrated without sedatives. Animal welfare activists have documented egregious examples of mistreatment of animals prior to slaughter, with chickens being smashed against walls and cattle having their throats slit and being hung by their legs without first being rendered unconscious.

●

In this era of global warming, researchers have cited the overall energy and pollution costs of different diets as an important reason to eat less meat. University of Chicago geophysicists Gidon Eshel and Pamela Martin calculate that it takes about 500 calories of fossil-fuel energy inputs to produce 100 calories' worth of chicken or milk; producing 100 calories' worth of grain-fed beef requires almost 1,600 calories. But producing 100 calories' worth of plant foods requires only 50 calories from fossil fuels. In terms of global warming, eating a typical American diet instead of an all-plant diet has a greater impact than driving a Toyota Camry instead of a gas-frugal Toyota Prius.[1] And that difference translates into an annual 430 million tons of carbon dioxide, 6 percent of the nation's total emissions of greenhouse gases.

Nutrition researchers in Germany have examined the ecological impacts of three kinds of diets: typical Western, low meat, and lacto-ovo vegetarian.[2] Compared to a typical diet, a low-meat diet uses 41 percent less energy and generates 37 percent less carbon dioxide equivalents (greenhouse gases) and 50 percent less sulfur dioxide equivalents (respiratory problems, acid

Greenhouse Gases

Global warming is occurring because increased amounts of carbon dioxide and other gases in the atmosphere trap extra heat and gradually warm our planet. While automobiles and fossil-fuel power plants are the biggest contributors to global warming, agriculture also plays a role.

● Livestock (mostly cattle) plus the manure lagoons on factory farms (mostly hog) generate an amount of methane that promotes about as much global warming as the release of carbon dioxide from 33 million automobiles. Methane is 23 times as potent as an equal amount of carbon dioxide.

● Nitrous oxide—which comes from degradation of manure and from fertilizer applied to cropland—is 300 times as potent as carbon dioxide in promoting global warming and accounts for 6 percent of the greenhouse effect in the lower atmosphere.

● Manufacturing fertilizer generates both carbon dioxide and nitrogen-containing greenhouse gases.

rain).* For a lacto-ovo vegetarian diet, the savings are even greater: 54 percent less energy, 52 percent less carbon dioxide equivalents, and 66 percent less sulfur dioxide equivalents.

Eating less meat and dairy products could greatly improve health, the environment, and animal welfare—especially if people replaced some of those foods with vegetables, beans, fruits, nuts, and whole grains (see "Changing Your Own Diet," p. 143). Most minimally processed plant foods are low in saturated fat and cholesterol and high in vitamins, minerals, and dietary fiber, and they are the only source of diverse phytonutrients. While producing more grains, vegetables, and fruits would require land, water, pesticides, and fertilizers, the amounts used would be small compared to the amounts saved by producing less animal-based foods. Even without cutting back on beef and dairy foods, just shifting the cattle industry away from feedlots and toward leaner grass-fed beef and getting the dairy industry to cut the saturated-fat content of milk would yield big dividends.

This pro-plant message, however, has one important caveat: Animal products do not have a monopoly on causing harm. Diets rich in salt, partially hydrogenated vegetable oils (with their trans fat), refined sugars, and refined flour also cause major health problems—heart disease, strokes, obe-

---

*Different gases have stronger or weaker effects on pollution. It is customary to convert them into equivalents of carbon dioxide and sulfur dioxide so their effects may be compared or combined.

sity, and tooth decay, to name a few. And certain crops—such as sugar cane in Florida and, indeed, almost any row crop grown in monoculture on large farms—wreak serious environmental damage.

While moving in a more vegetarian direction offers many benefits to health and the environment, a more omnivorous option is advocated eloquently by University of California journalism professor Michael Pollan in his recent book, *The Omnivore's Dilemma*. Pollan describes the multiple virtues of small farms that humanely and ecologically raise cattle, pigs, and chickens on pastures and in woodlands and sell their meat, milk, and eggs locally.[3] There's little room for factory farms, Wal-Marts, or Burger Kings in that vision, though the consumption of animal products could be at unhealthy levels. A more (or totally) plant-based diet could be as compatible with sustainable agriculture as diets that include animal products, but comparing the two approaches is a good reminder that no path is perfect: Each has its own compromises related to taste, cost, convenience, cultural values, health, ecology, animal welfare, and the vitality of rural America. Ultimately, what you eat is your choice.

Despite the well-recognized benefits of diets higher in healthy plant-based foods and lower in animal products (especially those produced on factory farms), relatively few people will change their diets (and few farmers will change their growing practices) without encouragement from new government policies. *Six Arguments*, therefore, suggests a range of policy options and programs (see "Changing Government Policies," p. 151). Some of our proposals would directly promote a healthier, more environmentally sound diet. Others might reduce consumption by increasing the price of animal products. And some would improve the lives of farm animals.

●

That's what *Six Arguments for a Greener Diet* is about. Now a few words about what it is *not* about. *Six Arguments* focuses on the United States, though the same logic applies to every other nation. The United States and other industrialized nations have largely passed through the "nutrition transition," meaning that diets that were once based largely on starchy grains and potatoes now include much greater amounts of meat. Hundreds of millions of people in India, China, Indonesia, and other developing nations are following our footsteps toward the meat counter. As Lester Brown, president of the Earth Policy Institute and a long-time analyst of global agriculture policies, has noted, the animal-rich American diet requires the production of four times as much grain per person as the average Indian diet.[4] If the entire world's population were to eat as much meat as Westerners, two-thirds more land would be needed than is currently farmed.[5] The increased demand for water, fertilizer, and pesticides and the concomitant increased pollution would be unsustainable and ultimately devastating to our planet.

Six Arguments for a Greener Diet puts the health, environmental, and animal welfare consequences of raising and eating livestock under the microscope, but does not delve into the whys and wherefores of the situation. Why are so many animals allowed to be raised in miserable conditions? Why are restaurants permitted to market fatty hamburgers and other unhealthy foods to young children? Why are livestock operations that raise thousands or tens of thousands of chickens, pigs, and cattle allowed to pollute waterways and the atmosphere with tons of smelly, drug-tainted manure and global-warming pollutants? Why are huge soybean and grain farms allowed to use such large amounts of fertilizer and pesticides that the run-off pollutes rivers, lakes, and even oceans? Why do farmers who grow crops to feed livestock receive billions of dollars in annual subsidies, hundreds

of times as much as fruit and vegetable growers receive? Why does the federal government not shape its farm and health policies around its sensible *Dietary Guidelines for Americans* and the vitality of rural communities?

It's questions like those that activate dozens of agribusiness, food industry, environmental, health, and consumer groups at the local, state, and national levels. The answers to the "why" questions are matters of politics, not science, and typically revolve around money and livelihoods. The makers of pesticides, fertilizer, and animal drugs; the cattle, hog, poultry, and dairy industries; the large grain companies and grain farmers—they all defend the status quo. They pour millions of dollars each year into campaign contributions, lobbyists' salaries, and advertising campaigns. They wine and dine politicians—often over fatty steaks—and use hardball tactics to rein in any rare elected official who dares stray from the proper path. (Senators will long remember how, in 1980, the cattle industry successfully campaigned to defeat South Dakota senator George McGovern because he dared recommend that people eat less beef.) And, by making use of the "revolving door," top officials from the cattle, pork, dairy, and other food- and agriculture-related industries become top officials in the U.S. Department of Agriculture, and many former legislators and Department of Agriculture officials enjoy more lucrative, and no less influential, careers on Washington's K Street, where they lobby for those industries.

Getting the "why" questions answered in a way that protects humans, animals, and the environment will require the involvement of thousands of concerned citizens, nonprofit organizations, concerned farmers and companies, legislators, and government officials at the local, state, and national levels. Considering how important these matters are, now is the time to start. Meanwhile, each of us can quietly do our part—in our kitchens, grocery stores, farmers' markets, and backyard gardens.

The Context

The Fatted Steer

G rain-fed beef. Since the 1950s, that term has conjured up thoughts of tender, juicy, delicious meat. Grain-fed beef is advertised by supermarkets and featured by restaurants. Omaha Steaks, a national retail and mail order company, proclaims: "We select the finest grain-fed beef for superior marbling, flavor and tenderness." Morton's, the high-end steakhouse chain, "serves only the finest USDA prime-aged, Midwest grain-fed beef." And the latest epicurean delicacy, Kobe beef—advertised as the "most flavorful and tender beef on the Planet"—is fed grain (and often beer).[1] The implication is that beef from cattle that were not grain-fed is tough, tasteless, and simply not worth eating.

In truth, grain-fed beef, which accounts for some 85 percent of American beef, epitomizes much of

- Grain-fed beef is rich in saturated fat and cholesterol, which promote heart disease.

- Growing corn and other crops for cattle feed requires enormous amounts of fertilizer, water, pesticides, land, and fossil fuel.

- Feedlot cattle eat a grain-rich diet that can cause digestive, hoof, and liver diseases and necessitates the continuous feeding of antibiotics.

- Grass-fed cattle are less harmful to the environment and provide leaner beef, but still generate air pollution. Any kind of beef increases the risk of colon cancer.

what is wrong with both the American "factory" approach to livestock production and the American diet. They eat a diet that sickens them. They generate air and water pollution. They pack on fattier meat. And, to top it off, grain-fed beef doesn't necessarily taste better than grass-fed beef.

A sensible argument for raising cattle and other ruminants is that their manure fertilizes grasslands, and they can convert into meat or milk the nutrient- and fiber-rich plant matter—grasses, cornstalks, and the like—that humans cannot digest. Raising cattle that way, though not without problems of its own, expands the food supply. However, in the United States, that rationale for including beef in the diet is undercut by the fact that the great majority of beef cattle spend months in feedlots eating grain, getting fat, and generating pollution.

The Objective: Cheap and Tender Beef

Restaurateur and former professional football player Dave Shula's "Views on Great Beef" include the note that "A great steak is all about flavorful, juicy and tender beef."[2] And an animal physiologist with the U.S. Department of Agriculture (USDA), discussing why he studies cattle proteins and genes, explains that "Tenderness is the most important trait to consumers."[3]

The cattle industry certainly wants to satisfy consumers' desires—and maximize its profits. Fortunately for the industry, techniques that produce tasty meat also turn out to be the cheapest way to raise cattle. The high-energy diets dished out at feedlots speed the animals' growth, with much of the increased weight taking the form of fat. Much like a restaurant that tries to "turn" its tables as quickly as possible, the faster growth rate gets the cattle to market sooner. So with both gastronomic and financial motives in place, cattle producers have adopted practices that yield a very fatted steer indeed.

Choosing to Produce Lean or Fatty Beef

For thousands of years, farmers have employed such factors as breeding and feed to shape the nature and yield of the meat (or milk or pork or chicken). In recent decades, scientists and agribusiness firms have turned the art of meat production into a science, with careful research supplanting happenstance.

Unfortunately, the practices that lead to the fastest production and cheapest prices are not what's best for the consumer's health.

They Are What Their Parents Are

Breed is a major determinant of cattle's fat content. Angus, Hereford, and crosses with other breeds are the most popular breeds in the United States, not least because they are among the fattiest. They have the largest amounts

Quality and Yield: Understanding USDA Meat Grades

Because fat content is important to beef purchasers, the U.S. Department of Agriculture has established a complex grading system that gives high grades to beef that is well-marbled with intramuscular fat.[4] About 80 percent of all beef cattle and cows are graded by visual inspection at the slaughterhouse. The fattiest meat (8 percent marbling or higher) rates as Prime, the next fattiest (5 to 7 percent marbling) as Choice, and the leanest meat (3 to 4 percent marbling) as Select.[*] In recent years, about 40 percent of cattle were graded as Select, 60 percent as Choice, and 2 to 3 percent as Prime.[5] Restaurants and supermarkets pay a premium for that fatty Prime meat. Producers also receive a premium for such special USDA grading programs as "Certified Angus Beef" or "Certified Hereford Beef," which are breeds that yield mostly high-Choice beef (see figure 1).[6]

"External" fat—that is, fat outside of the edible beef used as steaks—is reflected in USDA's "yield grades." The lower the grade on a scale of 1 to 5, the less fat.[7] Of meat that is graded, 85 percent is USDA yield grade 2 or 3. Although some producers argue that the quantity of external fat is unimportant because most of it is trimmed from beef cuts, much of that fat eventually ends up back in the food supply when it is blended with lean ground beef or used as shortening in baked goods.[8]

---

[*]An even leaner grade of beef, Standard, represents only 0.3 percent of all meat that is graded.

of external fat and the highest marbling scores, and they provide the highest percentages of Choice meat (see figure 1). The Limousin and Chianina breeds are far leaner. In Italy, in fact, the Chianina breed is prized for its lean meat. In the United States, it is often crossbred with other cattle—such as the Hereford—to increase marbling in the Chianina or decrease back fat in the Hereford.

They Are What They Eat

What cattle are fed greatly influences how fatty their meat will be. In a study at Ontario's University of Guelph, Ira Mandell and his colleagues let Limousin calves graze for eight months.[9] The cattle were then fed either grain or mostly alfalfa hay for seven months (see table 1). The average carcass weight of the grain-fed steers was 45 pounds more than that of the hay-fed steers, reflecting faster growth on a high-energy diet. The layer of back fat over the longissimus muscle (the main muscle in rib and strip loin cuts) was twice as thick in the grain-fed steers. And meat from the grain-fed steers contained almost twice as much intramuscular fat. The hay-fed steers, on the other hand, produced more lean meat than their grain-fed counterparts.

Figure 1. **Percentage of fattier meat in selected cattle breeds**[10]

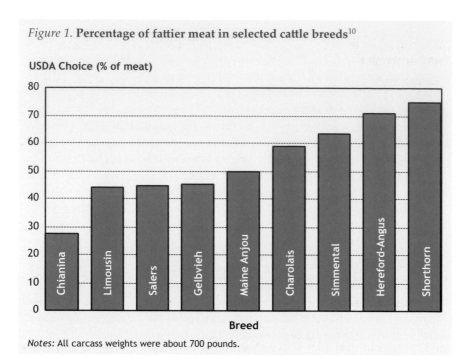

USDA Choice (% of meat)

Notes: All carcass weights were about 700 pounds.

While some breeds are inherently higher in fat, they will be leaner if they graze on pasture. In a study conducted at North Carolina State University, Angus steers were kept on pasture or fed corn until they weighed about 1,200 pounds (see table 2).[11] The grass-fed steers took about 1½ months longer to reach that weight, and their meat contained much less fat marbling than that from the grain-fed steers: Grass-fed beef was on the lean side of USDA Select, while grain-fed beef was on the high side of USDA Choice. Although the average carcass weight of the grass-fed steers was 75 pounds less than that of the grain-fed steers, the area of their longissimus muscle was almost as large as that of the grain-fed steers—a sign that grass-fed cattle can yield almost as much edible meat as grain-fed cattle. Moreover,

Table 1. **Carcass traits of Limousin steers fed grain or hay for 209 days**[12]

Carcass trait	Grain-fed steers	Grass-fed steers
Carcass weight	720 lb	674 lb
Total fat	27%	19%
Intramuscular fat	4.0%	2.7%
Back fat over longissimus muscle at slaughter	0.4 in	0.2 in
Lean meat	395 lb	409 lb

Table 2. **Carcass traits of Angus steers fed grain or grass and slaughtered at similar weights**[13]

Carcass trait	Grain-fed steers	Grass-fed steers
Days on diet	91	133
Weight at beginning of experiment	896 lb	909 lb
Slaughter weight	1,260 lb	1,190 lb
Carcass weight	750 lb	675 lb
Marbling score*	6.2	4.5
USDA quality grade[†]	17.5	15
USDA yield grade[‡]	3	2.2
Longissimus muscle area	13.1 sq in	11.9 sq in

* Scoring system designed by researchers to match USDA's scoring system: 4 = slight degree of marbling; 5 = small; 6 = modest; 7 = high.

[†] Scoring system designed by researchers to match USDA's scoring system: 16 – Select; 17 = Choice; 18 = High Choice.

[‡] Yield grade is measured on a scale of 1 to 5, with 5 containing the highest amount of waste fat.

the lower yield grade indicates that the grass-fed beef had less low-value external fat.

An animal's diet can override the effect of breed. Feeding grain to a leaner breed of cattle over longer periods can result in meat that is as fatty as that produced by a fattier breed. The University of Guelph researchers compared the Red Angus breed with the leaner Simmental.[14] Both groups of animals were finished with a high-grain diet and slaughtered when they reached the same back-fat thickness (about 0.4 inches, determined by ultrasound). The Simmental took about 2½ months longer than the Red Angus to reach the same amount of back fat and, thus, spent substantially more time on feed grains. The Simmental outweighed the Red Angus at slaughter by 45 pounds, and, despite its "lean" reputation, had a slightly higher marbling score and total (external and internal) fat content. So, just because meat comes from a normally lean breed does not automatically mean that the meat is lean.

Younger Is Leaner

The age at which cattle are slaughtered strongly affects fat content. In a study led by Susan Duckett at the Oklahoma State University Meat Laboratory, grain-fed Hereford-Angus cattle were slaughtered after 28-day intervals on high-energy diets.[15] After periods longer than 84 days, cattle progressively accumulated wasteful, external fat without increases in the palatability (taste, juiciness, and tenderness) of their meat. Between 84 and

112 days on feed grains, the cattle experienced the largest increase in external fat and marbling. During that period, the content of intramuscular fat more than doubled, moving the meat from USDA Select to Choice. Those results suggest that limiting grain feeding to 84 days—many cattle are on feed for up to 190 days—could provide much more healthful meat.

Fatty Meat Clogs Arteries...

Fattening cattle on grain is the quickest way to get them to market, but the higher fat content of feedlot beef is life threatening. Beef is a major source of saturated fat and cholesterol, which increase levels of the harmful kind of cholesterol in our blood. That clogs arteries and increases the risk of heart attacks, the nation's number-one cause of death. While consumers can easily cut away the outside fat on steaks, they can't remove the fat that marbles steaks or the fat in hamburgers and meatloaf.

Grass-fed beef is usually lower in fat and less conducive to heart disease.[16] But, as we will discover in the next chapter, any kind of beef—especially processed meats such as sausages—promotes colon cancer.

...And Doesn't Necessarily Taste Better

Americans have been trained to salivate at the mention of grain-fed beef. "This creates well-marbled, tender, flavorful steaks. Marbling is the easiest way to spot a high quality steak," says Iowa Corn Fed, a mail-order company that charges as much as $35 a pound for a steak.[17] One study found that pasture-raised beef sometimes has a "grassy" off-flavor. A University of Nebraska study found that half the taste testers preferred corn-fed beef, but the other half either preferred Argentinian grass-fed beef or were undecided.[18]

Taste experts agree that corn-fed beef tastes *different* from grass-fed beef, but not necessarily better. Corby Kummer, food editor for the *Atlantic Monthly*, says "Grass-fed beef tastes better than corn-fed beef: meatier, purer, far less fatty, the way we imagine beef tasted before feedlots and farm subsidies changed ranchers and cattle."[19] Careful, moist cooking, such as using marinades, helps reduce any stringiness.

Many studies dispute Kummer, presumably because taste is subjective and tasters bring with them their expectations of what tastes good.[20] But some of the studies make a case for grass-fed beef. Mandell and his colleagues at the University of Guelph compared meat from the popular Hereford breed and the leaner Simmental breed. Cattle of each breed were fed mostly grass or mostly grain. A trained taste panel judged meat from both breeds—whether they ate grain or grass—to be equally palatable.[21]

Another study—sponsored in part by the National Cattlemen's Beef Association—found that among top loin, top sirloin, and top round steaks, consumers showed barely any preference for the fattier Choice grade over Select.[22] The study, conducted by Texas A&M University researchers, found that the more often consumers purchased leaner meat, the less able they were to distinguish among quality grades. They concluded that the "USDA quality grade may be limited" in indicating the taste of a steak. Taste is more culturally determined than genetically determined. It's no

surprise, then, that people prefer the kinds of beef they grew up with: fatty grain-fed in the United States and lean grass-fed in Argentina (the biggest beef-consuming country in the world). But we suspect that many more consumers would enjoy grass-fed beef if they both tasted it and were told of its health and environmental advantages.

Although beef production is geared to delivering fattier Choice or Prime meat, some health-conscious consumers are seeking leaner meat. Some companies, such as Laura's Lean Beef, pay ranchers a premium for cattle that yield leaner Select grade beef. Other ranchers, such as Maverick Ranch and Coleman, market grass-fed or organic beef, which is often leaner than regular beef, and are getting a premium for it. For example, Hawthorne Valley Farms, which boasts several hundred acres of lush pastureland, charges up to $20 per pound for grass-fed tenderloin steaks at local farmers' markets.[23] In response to this growing consumer demand, even the Cattlemen's Beef Board sometimes highlights the low fat content of certain steaks.[24]

Raising Cattle Harms the Environment...

Raising tens of millions of cattle not only provides meat that promotes heart disease and sometimes causes food poisoning (see Arguments #1

Grass-Fed Beef: Better, but Not a Health Food

Grass-fed beef is typically leaner than feedlot beef, a major advantage; and grazing on pasture spares the need for about 5,000 pounds of grain per animal. Beyond that, some advocates maintain that grass-fed beef is rich in two special kinds of fat—conjugated linoleic acid (CLA) and omega-3 fatty acids—that confer health benefits. One purveyor, American Grass Fed Beef, emphasizes that its "grass fed beef is high in heart friendly essential fatty acids."[25] As yet, however, the evidence for such benefits is scanty, and even lean beef modestly increases the risk of heart disease and promotes colon cancer.

Conjugated Linoleic Acid

In the early 1980s, scientists suggested that CLA in beef might help fight obesity and prevent cancer. However, studies over the past two decades generally have been unsuccessful in linking the consumption of grass-fed beef to those "near-magical" (as one skeptical scientist stated) results.

- *Weight gain.* Michael Pariza—the University of Wisconsin scientist who first identified CLA in beef and heralded its possible benefits—found that CLA reduces weight gain in laboratory mice, with possibly smaller benefits in other lab animals.[26] However, Pariza notes that the fat mostly reduces future weight gain, not the initial weight. An industry-sponsored study suggests that CLA might lower the percentage of body fat, but not weight.[27] An added complexity is that meat and dairy products contain one form of CLA, while dietary supplements contain an additional form. Only the form in supplements affects weight in animals.

    The bottom line is that human studies have not shown a benefit,[28] and some research indicates that supplements may increase the risk of diabetes, heart disease, and other problems.[29] In 2002, the Institute of Medicine, a part of the

and #2), but also wreaks environmental havoc, as detailed in Arguments #3, #4, and #5. A mid-sized feedlot with 10,000 cattle churns out half a million pounds of manure each day—equivalent to a city such as Washington, D.C., with 500,000 residents. That mountain of fragrant manure pollutes the air and sometimes pollutes streams and rivers, killing plants and animals. The methane that cattle and their manure produce has a global-warming effect equal to that of 33 million automobiles.

National Academy of Sciences, stated that "research on the effects of CLA on body composition in humans has provided conflicting results" and declined to set a recommended intake level.[30] Overweight individuals should run—but not to grocery stores for grass-fed beef or drug stores for supplements.

● *Cancer.* When female rats predisposed to mammary (breast) tumors were fed a diet containing 0.5 percent to 1 percent CLA, existing tumors grew more slowly or stopped growing, and fewer new tumors developed. Also, the tumors did not spread to other organs.[31] In 1989, *USA Today* opined that beef "aids [the] war on cancer" and could "be made into a drug" if CLA proved beneficial to humans.[32] But the Institute of Medicine threw cold water on that notion, too, saying that "to date, there are insufficient data in humans to recommend a level of CLA at which beneficial health effects may occur."[33] Even if beef's CLA turns out to protect against cancer, grass-fed beef's lower fat content—its real health advantage—would reduce the benefits from the higher content of CLA in its fat.[34]

Overall, the evidence that CLA offers health benefits is skimpy. And if CLA ever were proven to offer benefits, doctors certainly would prescribe pills, not burgers.

Omega-3 Fatty Acids

Some people claim that grass-fed beef is especially healthful because it contains about five times as much omega-3 fatty acids as grain-fed beef.[35] Those are the same fatty acids—eicosapentaenoic acid (EPA) and docosahexaenoic acid (DHA)—that are found in fish oil and appear to prevent heart attacks and possibly strokes.[36] Beef also contains small amounts of alpha-linolenic acid, some of which the body can convert to EPA and DHA.[37] But the amounts of all of those fatty acids are small.

The American Heart Association recommends that people without heart disease eat fish twice a week, as well as flaxseed, canola, and soybean oils. People with heart disease should consume about 1 gram of EPA and DHA per day.[38] To get that amount from grass-fed beef would mean eating about 5 to 10 pounds of rib steaks.[39] Clearly, fish and dietary supplements are better sources: Three ounces of bluefin tuna provide 1.5 grams of the fatty acids; 3 ounces of Atlantic salmon provide 1.9 grams.[40]

Feeding grain to cattle makes a bad situation worse. It takes about 7 pounds of corn to put on 1 pound of weight. That's why over 200 million acres of land are devoted to producing grains, oilseeds, pasture, and hay for livestock.[41] Moreover, cultivation of those crops requires 181 million pounds of pesticides, 22 billion pounds of fertilizer, and 17 trillion gallons of irrigation water per year. The fertilizer and pesticides pollute the air, water, and soil, while irrigation depletes natural aquifers built up over millennia.

Grazing's Pluses and Minuses

Grazing is better in many ways than feeding grain to cattle, but it still exacts environmental costs. Cattle that eat grass and roughage release more methane (a gas that causes global warming and is 23 times more potent than carbon dioxide) than cattle on a high-energy feedlot diet, because grass-fed cattle take about 10 to 20 percent longer to reach market weight.[42] Those longer lives also mean more manure—about 3,500 to 5,000 pounds per animal (60 pounds per day). That manure, though, is dispersed widely on pastureland, enriching the soil and nourishing the growth of plant life.[43]

...And the Cattle, Too

One measure of our humanity is how well we treat animals. While pets, of course, are often pampered almost like children, livestock are another story. Aside from sometimes being branded with a burning hot iron and, in the case of males, castrated without the benefit of sedation or painkillers, beef cattle have a pretty good life for their first year or so, living on the range. But then virtually all cattle are shipped in crowded trucks—exposed to the elements and banged about—to feedlots, where they dwell for up to six months in manure-befouled pens and eat a high-energy corn-based diet that sometimes causes liver, hoof, and gastrointestinal illnesses and occasionally even fatal bloating. (Shipping the animals to the feed is cheaper than hauling the feed to them. Indeed, in the case of chickens, corn and soybean meal account for 60 percent of the cost of production.[44])

When they've reached market weight, feedlot cattle (along with small numbers of pasture-raised cattle) are shipped for the final time to a slaughterhouse where they have a small, but real, risk of a slow, painful death. From that point on, the cattle exact a sort of posthumous revenge: First to suffer are the workers in slaughterhouses and meat processing plants who experience everything from repetitive movement injuries to knife wounds. Next are the unwitting consumers, who may suffer foodborne illness in the short term or fatal heart attacks in the long term.

What It All Means

Raising cattle provides valuable nutrients, leather, and by-products used by the food and drug and other industries. But considering how most cattle are raised, those positives are outweighed by a host of negatives. To protect our own health and our country's environment, the best thing we could do would be to eat less, leaner, or no beef. Should that happen on a large enough scale, vast areas of cropland could be freed up, allowing the land to regain much of its original fertility and biodiversity or to be planted in more healthful fruit and vegetable crops or crops that would provide biofuel.

But as long as people *do* eat beef, raising cattle on pastureland—instead of feeding them grain—would dramatically reduce the fat content of beef, the waste and pollution of water and the fouling of air caused by manure and agricultural chemicals, and the misery experienced by the cattle consigned to feedlots.

The Arguments

Argument #1.

Less Chronic Disease and Better Overall Health

Our Diet Is Killing Us

At least one of every six deaths in the United States—upwards of 340,000 each year—is linked to a poor diet and sedentary lifestyle.[1] The average American is about as likely to die from a disease related to diet and physical inactivity as from smoking tobacco—and far likelier to die from diet and inactivity than from an automobile accident, homicide, or infectious disease such as pneumonia.[2] Among nonsmokers, the combination of diet and physical inactivity is the *single* largest cause of death.

The specific diet-related diseases that fell so many of us include heart disease, certain cancers, stroke, and diabetes. Those and other chronic diseases (so called because they develop

- The saturated fat and cholesterol in beef, pork, dairy foods, poultry, and eggs cause about 63,000 fatal heart attacks annually.

- Less than a quarter of all adults eat the recommended number of daily servings of fruits and vegetables—foods that reduce the risk of heart disease and cancer.

- Vegetarians enjoy lower levels of blood cholesterol, less obesity, less hypertension, and fewer other problems than people whose diet includes meat.

and progress over many years) are caused in part by diets too poor in healthy plant-based foods and too rich in unhealthy animal-based foods.

We Eat Too Much of What's Bad for Us...

Obesity, which is directly linked to diet and a sedentary lifestyle, markedly increases a person's risk of heart disease, hypertension (high blood pressure), diabetes, and some cancers. Rates of obesity have doubled in children and adults and tripled in teenagers since the late 1970s, which is not surprising, since—thanks to ubiqui-

tous high-calorie foods—the average adult eats 100 to 500 calories more per day and—thanks to modern conveniences—exercises less.[3] The additional calories have come mainly from the least healthy foods: white flour, added fats and oils, and refined sugars.[4]

Moreover, Americans are eating more flesh foods—beef, pork, chicken, turkey, and seafood. In 2003, for instance, Americans ate more of each of those foods than they did a half-century earlier (see figure 1 and table 1). Fortunately, the biggest increase was for poultry, which is not directly linked to chronic disease. However, a lot of that chicken—and fish too—is not baked or grilled, but deep fried in partially hydrogenated oil. That oil contains trans fat, one of the most potent causes of heart disease. Meanwhile, Americans cut their consumption of beef by 33 percent since 1976; that is likely due both to health concerns and lower chicken prices.

Our inconsistent efforts to eat healthy diets extend to non-meat foods as well. Although we are eating one-third fewer eggs—the yolks of which are our biggest source of cholesterol and thus contribute to heart disease—than we did in 1953, we are eating four times as much cheese—which is high in saturated fat and promotes heart disease (see table 1).

Figure 1. **Major sources of animal protein produced in the United States**[5]

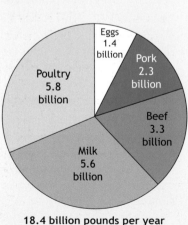

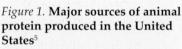

18.4 billion pounds per year

Table 1. **Per capita availability of major sources of meat, poultry, and seafood; dairy foods; and eggs**[6]

Year	Beef & veal	Pork	Chicken	Turkey	Fish & shellfish	Milk & yogurt	Cheese	Eggs
1909	56	41	10	1	10*	34	4	293
1953	61	39	15	4	11	37	7	379
1976	92	41	29	7	13	30	16	270
2003	62	49	58	14	16	23	31	253

Notes: Figures for meat, poultry, and seafood represent the numbers of trimmed (edible) pounds per capita that were available in the food supply; the remaining figures represent the per capita numbers of gallons (milk and yogurt), pounds (cheese), or eggs that were available in the food supply. Due to waste and spoilage, actual consumption is lower. Beef consumption peaked in 1976.

*Figure is for 1929, the first year for which data are available.

Looking at other non-animal-derived portions of our diet, we are consuming massive amounts of nutritionally poor plant-based foods, notably:

- refined grains (white bread, white pasta, and white rice), which are stripped of much of their nutrients and dietary fiber;
- soft drinks and other foods high in refined sugars (including high-fructose corn syrup), which replace more healthful foods and promote obesity; and
- baked goods and fried foods made with partially hydrogenated vegetable oil and palm, palm kernel, and coconut oils, which promote heart disease.

Finally, there's salt. The large amounts of salt in most packaged and restaurant foods and processed meats increase blood pressure, which increases the risk of heart attacks and strokes.

...And Not Enough Whole Grains, Fruits, and Vegetables

The U.S. Department of Agriculture (USDA) estimates that the average adult eats only one serving of whole grains daily.[7] In contrast, the *Dietary Guidelines for Americans* recommends that at least half of our 6 to 10 daily grain servings should be whole grain.[8] The

The Cardiovascular Benefit of Eating Less Meat and Dairy

Probably the biggest health benefit from eating less animal products (other than fish) is a lower risk of heart disease. The Center for Science in the Public Interest estimated the approximate benefit based on the:

- amounts of different fatty acids and cholesterol that are supplied by various animal products,
- impact of saturated fat and cholesterol on blood cholesterol levels, and
- relationship between blood cholesterol and heart disease.

We first estimated how our consumption of fats and cholesterol would change if all the beef, pork, milk and cheese, poultry, and eggs were removed from the average diet and either not replaced or replaced with foods that did not affect the risk of heart disease.[9] Next, we projected how those changes in fat and cholesterol intake would affect blood cholesterol levels by averaging the results from formulas developed by several leading researchers.[10] We then assumed that a 1 percent increase in blood cholesterol—total or low-density lipoprotein (LDL, or "bad" cholesterol) increases heart disease mortality by 2 percent.[11]

Those calculations indicate that avoiding animal fats would save about 63,000 lives per year (see figure).[12] Because that estimate is based on inexact assumptions, the true total might easily be 25,000 more or fewer lives per year. The number of lives saved would be dramatically greater if one assumed that people replaced much of the meat and dairy products with healthier plant-based foods or fish. The economic benefit of avoiding the fat would be about $100 billion

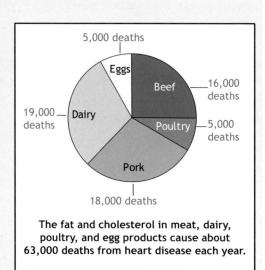

The fat and cholesterol in meat, dairy, poultry, and egg products cause about 63,000 deaths from heart disease each year.

a year or in excess of $1 trillion over 20 years.[13] On the other hand, the same methodology indicates that the healthy unsaturated fats in salad oils currently save about 7,000 lives a year.

Of course, we could reap some of those benefits by switching to lower-fat animal products—such as from beef to chicken or even buffalo and to low-fat dairy foods.

The Economic Benefits of a More Plant-Based Diet

Diseases related to a diet too poor in plant foods and too rich in animal foods contribute to skyrocketing health-care costs. The annual cost of angioplasties and coronary bypass operations is about $50 billion, with statin heart-disease drugs adding $15 billion.[14] Spending to treat high blood pressure (including $15 billion for drugs[15]), stroke, diabetes (another $7 billion for drugs), and cancer add additional billions.[16] And, of course, on top of the medical costs are the incalculable amounts of pain and suffering (of both the people with the diseases and their friends and relatives) and lost productivity.

Eating a more plant-based diet wouldn't eliminate all those costs, but would certainly move us well along in the right direction. One study estimated that going vegetarian would save the nation $39 billion to $84 billion annually.[17] If obesity—which is much less common in vegetarians than others—were eliminated, we could save about $73 billion a year.[18]

USDA also estimates that we are eating 1.2 servings of fruit and 3.7 servings of vegetables per day, considerably less than the recommended 5 to 10 daily servings.[19] And, disappointingly, potato chips and French fries (which are often cooked in partially hydrogenated shortening) here count as "vegetables." Indeed, one-third of the vegetables that we eat are iceberg lettuce and potatoes, two of the least nutritious. We are consuming only one-third the recommended amount of the most nutritious vegetables: deep yellow and dark leafy green vegetables, and beans.[20]

According to the USDA, we're very slowly increasing our consumption of vegetables: Fresh vegetables are up 33 percent, and total vegetables are up 25 percent, since 1970. Surprisingly, though, fruit consumption is up only 12 percent over that period and has not increased at all in 20 years.[21]

As our diets have been buffeted by cultural, economic, and other factors, the evidence that certain dietary changes can reduce our risk of chronic disease has become much stronger. Much of the research shows that people who eat more plant-based diets, such as those traditionally eaten in Mediterranean or Asian countries, are generally healthier than those eating the typical American, Canadian, or northern European diet.

How Do We Know?

Study after study points to meat and dairy products, especially fatty ones, as causes of chronic diseases. The harm results both from specific constituents in animal products (such as saturated fat and cholesterol) and from pushing healthier nutrient-rich plant foods out of the diet. This section presents the science behind the (by now) commonly accepted premise that eating too many of the wrong animal products and too few of the healthiest plant foods does tremendous harm to our health. Again, a common-sense caveat: Modest amounts of fatty fish and low-fat dairy, meat, and poultry products—even an occasional hot dog or cheeseburger—certainly can fit into a healthy diet. The problems arise from immoderation.

One approach to understanding the influence of diet on health is to compare groups of people who eat very different diets. Such "observational" studies can provide important insights into what constitutes a health-promoting diet, though they cannot determine with certainty the particular elements in the diets—or other aspects of the subjects' lives—that are responsible for the better health. We review those studies first, then examine "intervention" studies, which are better able to identify causes and effects. Finally, we examine the health effects of specific foods and nutrients.

Observational Studies Show That Vegetarians Live Longer and Are Less Prone to Chronic Diseases

Studies that compare disease patterns in people with different kinds of diets help identify factors that cause or prevent diseases. For example, dif-

ferences in disease rates between vegetarians (or vegans, who abstain from all animal products, including dairy and eggs) and non-vegetarians can help identify the effects of meat and other animal products. The weakness of this "observational" approach is that factors other than diet—such as physical activity, air pollution, use of legal and illegal drugs, and cigarette smoking—affect dis-

Meatless meals offer an incredible variety of tastes, textures, and smells.

ease rates as well. Scientists try to account for those kinds of factors, but it is impossible to know about and account for everything.

Seventh-day Adventists Eat a More Plant-Based Diet and Live Longer and Healthier Lives

Seventh-day Adventists (SDAs), whose religion advocates abstinence from meat and poultry as well as alcohol and tobacco, have provided invaluable evidence on lifestyle and health.[22] About half of American SDAs follow a vegetarian diet or eat meat less than once a week. About one-quarter of SDAs follow a meatless lacto-ovo vegetarian diet, which includes dairy products and eggs, and about 3 percent are vegan. Generally, even non-vegetarian SDAs eat less meat than does the average American. Vegetarian or not, SDAs also tend to be physically active and eschew tobacco and alcohol. So, by comparing vegetarian and non-vegetarian SDAs and adjusting for factors such as smoking, physical activity, and alcohol, the effects of a vegetarian diet can be teased out. Vegetarian SDAs may also be compared to the general population to shed light on the health effects of a lacto-ovo vegetarian diet.

SDAs, on average, consume less saturated fat and cholesterol and more dietary fiber than the average American.[23] They eat more fruit, green salads, whole wheat bread, and margarine and less meat, cream, coffee, butter, and white bread. The same is true of vegetarian SDAs compared to non-vegetarian SDAs.[24]

Key findings from studies of SDAs include the following:

- *Longevity.* Vegetarian SDA women live 2.5 years longer than non-vegetarian SDA women; vegetarian SDA men live 3.2 years longer than their non-vegetarian counterparts.[25]
- *Heart attacks.* Non-vegetarian SDA men have twice the rate of fatal heart attacks as vegetarian SDA men.[26] Similarly, the risk of fatal heart disease is more than twice as high for men who eat beef more than three times a week as for vegetarians.[27] However, beef consumption or vegetarianism does not clearly affect the risk of heart disease in women.[28]
- *Stroke.* SDAs in the Netherlands have about a 45 percent lower death rate from strokes than the total Dutch population.[29]
- *Cholesterol.* Among African American SDAs, LDL ("bad") cholesterol and triglycerides (the most common fat found in blood) were lower in vegans than in lacto-ovo vegetarians.[30] Both of those fatty substances promote heart attacks.
- *Hypertension.* Hypertension, which increases the risk of heart attacks and strokes, is twice as common in non-vegetarian SDAs as in vegetarians; semi-vegetarians (those who eat fish and poultry less than once a week) had intermediate rates.[31] Those findings apply to both men and women. When hypertension was defined as "taking antihypertensive

medication" (those with more severe hypertension), non-vegetarians had almost three times the rate of hypertension as vegetarians.[32]

● *Diabetes.* Diabetes is twice as common in non-vegetarian SDAs, whether male or female, as in vegetarians, with semi-vegetarians having an intermediate prevalence.[33]

● *Cancer.* Prostate cancer is 54 percent, and colon cancer is 88 percent, more common in non-vegetarian than in vegetarian SDAs.[34]

Some of those health benefits may be due not to particular nutrients in plant foods, but to the fact that bulky plant-based diets help reduce body weight. For example, for the average 5'10" male SDA, non-vegetarians weigh an average of 14 pounds more than vegetarians. For 5'4" female SDAs, non-vegetarians weigh 12 pounds more than vegetarians.[35]

Vegetarians Have Less Heart Disease, Hypertension, and Diabetes

Studies of non-SDA vegetarians yield similar results. For example, the USDA's 1994–95 Continuing Survey of Food Intake by Individuals asked more than 13,000 people whether they considered themselves to be vegetarian.[36] Self-defined vegetarians whose diets did not include meat made up 0.9 percent of this nationally representative sample. Compared to non-vegetarians, the self-defined vegetarians tended to consume less fat, saturated fat, and cholesterol and more fiber. Self-defined vegetarians also ate more grains, legumes, vegetables, and fruit. In addition, they consumed fewer calories and had lower BMIs (body mass index, which combines height and weight) than non-vegetarians.[37]

Several large studies in Europe have examined the health of vegetarians. The European Prospective Investigation into Cancer and Nutrition (EPIC) is an ongoing study involving over 500,000 people in 10 countries. The part of that study being conducted in the United Kingdom (EPIC-Oxford) involves more than 34,000 non-vegetarians and close to 33,000 non-meat-eaters (including people who eat fish, lacto-ovo vegetarians, and vegans).[38] Another British study, the Oxford Vegetarian Study, compared 6,000 vegetarians to 5,000 non-vegetarians.[39] (More than half of the non-vegetarian subjects in that study did not eat meat daily and, therefore, were not typical of the general British population.) Findings from those studies and similar ones include the following:

● *Cholesterol.* Vegans have 28 percent lower LDL cholesterol levels than meat-eaters. Lacto-ovo vegetarians and fish-eaters have levels between those of vegans and meat-eaters.[40] Based on blood cholesterol levels, the researchers estimated that heart disease rates would be 24 percent lower in lifelong vegetarians and 57 percent lower in lifelong vegans than in meat-eaters.

- *Heart disease.* Vegetarians have a 28 percent lower death rate from heart disease than meat-eaters.[41]
- *Blood pressure.* Vegetarians have lower blood pressure and a lower rate of hypertension than non-vegetarians. Vegans have the lowest blood pressure and the least hypertension, followed by vegetarians and fish-eaters; non-vegetarians have the highest rates of hypertension.[42] (Differences in body weight were responsible for about half of the variation in blood pressure; alcohol consumption and vigorous exercise accounted for some of the variation in men.[43]) The EPIC-Oxford study found hypertension rates of 9 percent in lacto-ovo vegetarians and 13 percent in non-vegetarians.[44]
- *Diabetes.* Mortality from diabetes is markedly lower for vegetarians (and for health-conscious non-vegetarians) than for the general population.[45]

As with the SDAs, some of the European vegetarians' health advantages are likely due to lower rates of obesity.[46] For instance, in the Oxford Vegetarian Study, overweight or obesity (BMI ≥ 25) was twice as common in non-vegetarian men, and 1½ times more common in non-vegetarian women, as in vegetarians.[47] In a Swedish study of middle-aged women, the risk of obesity was 65 percent lower in vegans, 46 percent lower in lacto-vegetarians (those who avoid meat, fish, poultry, and eggs), and 48 percent lower in semi-vegetarians compared to non-vegetarians.[48] On average, vegetarians are leaner than their non vegetarian counterparts by about 1 BMI unit

Meta-Analysis Find Vegetarians Have Less Heart Disease

Meta-analysis is a powerful statistical technique that combines the results from a number of similar studies into a single, large analysis. If done properly, such an analysis can provide more conclusive results than any single study. A meta-analysis of five studies (the Adventist Mortality Study, Health Food Shoppers Study, Adventist Health Study, Heidelberg Study, and Oxford Vegetarian Study) included a total of 76,172 vegetarians (both lacto-ovo vegetarians and vegans) and non-vegetarians with similar lifestyles.[49] The vegetarians had a 24 percent lower rate of fatal heart attacks than non-vegetarians. When compared to people who ate meat at least weekly, mortality from heart disease was 20 percent lower in occasional meat-eaters, 34 percent lower in those who ate fish but not meat, 34 percent lower in lacto-ovo vegetarians, and 26 percent lower in vegans. (The data on vegans may not be reliable, because the meta-analysis included only 753 vegans.) The meta-analysis did not find any difference in death rates from stroke or cancer between the vegetarians and non-vegetarians.

(roughly 6 pounds).[50] Differences in rates of obesity and BMI may be due to vegetarians' higher intake of fiber and lower intake of animal fat, although other unknown factors also appear to be involved.[51]

In sum, several large studies have found that vegetarians enjoy lower risks of major chronic diseases and longer lives than non-vegetarians. That is not surprising, considering that vegetarians have lower rates of obesity, lower saturated fat and cholesterol intakes, higher fiber intakes, and lower total and LDL cholesterol levels. Vegetarians' somewhat greater physical activity also plays a role. Smoking clearly is an important risk factor, but most recent studies adjust for it, as well as for age, alcohol use, and other readily identified factors. It is always possible, of course, that vegetarians may differ from other people in ways not accounted for in the studies.

Though the numbers of vegans in the studies are small, they tend to have lower serum total and LDL cholesterol, less hypertension, and a lower prevalence of obesity than lacto-ovo vegetarians. However, there is no evidence that vegans live longer than lacto-ovo vegetarians and semi-vegetarians.[52]

Followers of a "Prudent" Diet Are Less Likely to Have Heart Disease

Other major studies have found important connections between dietary patterns and heart disease. The ongoing Nurses' Health Study, which is managed by the Harvard School of Public Health, compared a "prudent" diet, with higher intakes of fruits, vegetables, legumes, whole grains, fish, and poultry, to the "Western" pattern, which is high in red and processed (sausage, bacon, and the like) meats, sweets, desserts, fried foods, and refined grains. After 12 years, among the more than 69,000 participants, the women who ate prudent diets were 36 percent less likely to develop heart disease than those who ate typical Western diets.[53] In a similar study of almost 45,000 male health professionals, a prudent diet was associated with about a 30 percent lower risk of developing heart disease or of dying from a heart attack.[54]

Intervention Studies Demonstrate Benefits of Low-Fat Vegetarian Diets

The bottom line from observational studies is that diets based more on plant foods—and that means carrots, not carrot cake—pay big health dividends. But the limitation of those studies is that vegetarians and other health-conscious individuals might be doing things besides eating more plant foods and fewer animal products that are the real reasons for their better health. Intervention studies overcome that limitation.

The best way to study the effect of diet on chronic disease is to assign participants randomly to two or more different diets. Such "intervention" studies include those in which subjects were placed on vegetarian or other kinds of diets, thus allowing researchers to evaluate the diets' relative strengths and weaknesses.

Low-Fat Vegetarian Diets Can Lower Blood Pressure and Decrease the Risk of Heart Disease

Vegetarian diets have proven to be remarkably beneficial for people who have cardiovascular disease. For instance, switching from ordinary omnivorous diets to a lacto-ovo vegetarian diet with similar sodium content but more fiber, calcium, and potassium reduced the blood pressure in subjects who had either normal or high blood pressure.[55] Differences in the kinds of fat, as well as the levels of minerals, in the vegetarian and non-vegetarian diets may have accounted for some of the differences in blood pressure.[56]

Several recent intervention studies examined the effect of a near-vegan diet high in phytosterols and soluble fiber on blood cholesterol levels.[57] Phytosterols are plant-based substances with a chemical structure related to cholesterol; they are added to some margarines, yogurts, and orange juice to reduce cholesterol absorption. The soluble fiber in such foods as oats, barley, psyllium, eggplant, and okra forms thick, sticky solutions that increase the excretion from the body of bile acids and lower blood cholesterol levels.

David Jenkins and colleagues at the University of Toronto placed people with high blood cholesterol levels on either (1) a near-vegan diet high in phytosterols, soluble fiber, and soy protein; (2) a low-saturated-fat lacto-ovo vegetarian diet; or (3) the latter diet along with a cholesterol-lowering statin drug. The diet that included phytosterols, soluble fiber, and soy protein improved cholesterol levels just as much as the lacto-ovo vegetarian diet plus the statin. Judging from the subjects' changes in cholesterol levels, blood pressure, and other measures, the near-vegan diet led to a 32 percent lower risk of heart disease than the lacto-ovo vegetarian diet. The near-vegan diet presumably had a greater effect because of the soluble fiber, phytosterols, and possibly soy protein (but see "Soy Foods: No Health Miracle," on p. 39). Jenkins notes, "There is hope

Morale-boosting communal dinners likely contribute to the success of the CHIP heart-health program (see next page).

that these diets may provide a non-pharmacologic treatment option for selected individuals at increased risk of cardiovascular disease."[58]

Based in part on the Toronto studies, the National Cholesterol Education Program, a part of the National Heart, Lung, and Blood Institute, recommended a combination of statins and dietary modifications for patients with high LDL cholesterol levels (above 130 milligrams per deciliter).[59]

Hans Diehl, a health educator at the Lifestyle Medical Institute in Loma Linda, California, has developed a community-based Coronary Health Improvement Project (CHIP) that involves hundreds of people at a time. CHIP encourages participants to switch to a near-vegan, low-fat diet (though most participants make more modest changes) and engage in walking or other physical activities.[60] After only a few weeks on the

The DASH and Mediterranean Diets

The Dietary Approaches to Stop Hypertension (DASH) intervention study used a more plant-based, but not vegetarian, diet. DASH examined the effects of a diet that includes twice the average daily consumption of fruits, vegetables, and low-fat dairy products; one-third the usual intake of red meat; half the typical use of fats, oils, and salad dressings; and one-quarter the typical number of unhealthy snacks and sweets. It emphasizes whole grains and severely limits salt (see "Changing Your Own Diet," p. 143, for more about this diet). Compared to a typical American diet, the DASH diet lowers blood cholesterol, blood pressure, and the risk of cardiovascular disease.[61] A major strength of this study was that the subjects were given all their meals, so the researchers knew exactly what they were eating.

A prominent French study, the Lyon Diet Heart Study, tested the effect on heart disease of a Mediterranean-type diet that emphasizes fruits, vegetables, bread and other grains, potatoes, beans, nuts, seeds, and olive oil and contains only modest amounts of animal products. In subjects who had already had a heart attack, the Mediterranean diet led to 50 to 70 percent fewer deaths, strokes, and other complications compared to those following a "prudent" Western-type diet.[62] Interestingly, blood cholesterol levels and cigarette use were similar in the two groups, indicating that other factors—possibly the threefold higher level of alpha-linolenic acid, an omega-3 fatty acid, in the experimental group—play important health roles. Also, weight loss was not responsible for the dramatic benefit—a finding unlike those in some other studies. Harvard Medical School professor Alexander Leaf commented that this "well-conducted" study showed that "relatively simple dietary changes achieved greater reductions in risk of all-cause and coronary heart disease mortality in a secondary prevention trial than any of the cholesterol-lowering [drug] studies to date."[63] He also noted that the subjects readily adhered to this diet.

program, participants typically eat more fruits and vegetables and less saturated fat and cholesterol than a control group. In one study, compared to the controls, the participants' average LDL cholesterol level declined by 14 percent.[64] Subjects who changed their diets also lost an average of 7½ pounds, and their rate of hypertension dropped in half. The CHIP study shows that a health-promotion program can provide enormous benefits to large groups of people in a cost-effective way.

Diet and Exercise Can Reverse Heart Disease

Dean Ornish, of the University of California in San Francisco, and his colleagues have done ground-breaking studies in patients with moderate to severe heart disease. The researchers prescribe a very-low-fat vegetarian diet (containing no animal products except nonfat dairy products and egg whites), along with moderate aerobic exercise, smoking cessation, and stress reduction. That regimen significantly improved cholesterol levels, at least temporarily. It also began unclogging arteries and preventing angina (the chest pain that occurs when the heart muscle does not get enough blood) and heart attacks.[65] Lipid-lowering statin drugs were not needed. The lifestyle changes were as effective as coronary bypass surgery in reducing angina. The subjects who ate the low-fat vegetarian diet and made other lifestyle changes lost an average of 24 pounds, which was undoubtedly an important factor in their improved health.

In another study by Ornish's research group, 440 men and women with coronary artery disease ate the same largely vegetarian diet and made the prescribed lifestyle changes.[66] After one year, the subjects enjoyed reduced blood lipids (13 percent lower LDL cholesterol in men, 16 percent lower in women), blood pressure (1 to 2 percent reduction in systolic blood pressure), and weight (5 percent in men, 7 percent in women).

Fighting Prostate Cancer with Lifestyle

Prostate cancer, which kills 30,000 American men each year, may be controlled with lifestyle changes, including a low-fat vegan diet. Dean Ornish and his colleagues at the University of California "treated" with diet, fish oil and other supplements, exercise, and other lifestyle changes half of a group of 93 volunteers with early prostate cancer. The other half received the usual care. After one year, prostate-specific antigen, one index of prostate cancer, decreased 4 percent in the treatment group but increased 6 percent in the control group. The cancer progressed sufficiently in six men in the control group, but in none in the experimental group, to warrant conventional medical therapy.[67]

Decades of eating fatty meat and dairy products can turn healthy arteries (like the opened and flattened human aorta at left) into ones afflicted with severe atherosclerosis (right).

In a smaller but much longer study, Caldwell Esselstyn of the Cleveland Clinic monitored 18 patients with severe coronary artery disease.[68] Most of them had suffered coronary problems after a previous bypass surgery or angioplasty. All of those who ate an almost entirely plant-based diet had no recurrence of coronary events over 12 years (a few patients took low doses of statin drugs some of the time). One patient who "fell off the wagon" had a heart attack and then resumed the program. The coronary arteries of 70 percent of the patients studied became less clogged. In Dr. Esselstyn's words, his patients had become "virtually heart-attack proof."

One concern about diets high in carbohydrates is that they tend to raise triglycerides and lower high-density lipoprotein (HDL, or "good" cholesterol), a prescription for heart disease. However, in China and Japan, where traditional diets are very high in carbohydrates, heart disease is almost nonexistent. That's probably because most Chinese and Japanese people have been lean and active—very different from the typical American. In addition, studies by Dean Ornish and David Jenkins of North Americans are reassuring. They found that diets high in carbohydrates from whole grains and beans, but low in white flour and sugar, led to major reductions in LDL cholesterol but had little or no effect on triglycerides and HDL cholesterol. The fact that Ornish's subjects were moderately active and lost weight undoubtedly helped. Ornish speculates that even when high-carbohydrate diets lower HDL cholesterol, that does not increase the risk of heart disease, while the low HDL cholesterol levels seen in people whose diets are high in refined sugars and starches do promote heart disease.[69]

A More Plant-Based Diet Can Treat Type 2 Diabetes

Low-fat vegetarian diets can treat type 2 diabetes, a terrible and increasingly common disease that causes everything from blindness to gangrene (and amputations) to heart disease. In one 26-day study of 652 people with diabetes, more than one-third of the insulin-using subjects who adopted a low-fat vegetarian diet were able to discontinue the insulin. Close to three-quarters of those on the vegetarian diet who were taking oral hypoglycemic medicines were able to stop taking them.[70] The vegetarian diet also yielded a 22 percent reduction in serum cholesterol and a 33 percent reduction in triglycerides. Some of those benefits were likely due to the subjects' losing an average of 8 pounds.

A study that combined a low-fat, high-fiber vegan diet with daily exercise and weight loss (11 pounds in 25 days) was also highly successful in treating type 2 diabetes.[71] The lifestyle changes eliminated the pain related to diabetes-caused nerve damage in most of the subjects. It also reduced fasting blood glucose levels, blood pressure, and the need for medications.

The results of intervention studies strongly indicate that a largely plant-based diet provides tremendous benefits—sometimes even as great as those achieved by powerful prescription drugs or surgery. Though some of those studies also involved relaxation, exercise, or low levels of drugs, diets consisting mostly of nutritious plant-based foods clearly are extremely effective at preventing or treating chronic diseases. The benefits include reductions in blood pressure, total and LDL cholesterol, blood glucose, clogging of arteries, and—most importantly—less cardiovascular disease and type 2 diabetes.

Building on that body of research, leading health agencies in the United States and abroad have developed quite similar dietary advice (see table 2). They stress the benefits from beans, whole grains, fruits, vegetables, and seafood, along with physical activity, and the harm that is associated with fatty meat and dairy products.

What Specific Foods Should We Be Eating—and Avoiding?

The studies we have discussed compared the health effects of widely different *diets*. Researchers also have studied the health benefits and risks of specific *food groups*, such as fruits and vegetables, and meat.

Fruits and Vegetables

Americans are eating slightly more fruits and vegetables today than the paltry amounts we ate 35 years ago, but still far less than the recommended 5 to 10 servings per day. Fruits and

Table 2. **Health experts' advice on diet, physical activity, and chronic disease**[72]

Disease	What increases risk	What decreases risk
Heart disease	• Saturated fat (especially meat and dairy)[DG, WHO] • Cholesterol (meat, dairy fat, egg yolks)[DG, WHO] • Trans fat[DG, WHO] • Salt[DG, WHO] • Obesity/overweight[DG, WHO] • Sedentary lifestyle[DG, WHO]	• Vegetables, fruits[DG, WHO] • Whole grains[DG, WHO] • Legumes[WHO] • Fish, fish oil[DG, WHO] • Fiber[DG, WHO] • Linoleic acid[WHO] • Alpha-linolenic acid[WHO] • Oleic acid[WHO] • Nuts (unsalted)[WHO] • Physical activity[DG, WHO] • Potassium[WHO] • Plant sterols/stanols[WHO] • Folate[WHO]
Cancer*	• Obesity/overweight[ACS, DG, WHO] • Alcohol[ACS, DG, WHO] • Meat (fresh and preserved)[ACS, WHO] • Dairy products (high-fat†)[ACS] • Sedentary lifestyle[DG]	• Vegetables, fruits[ACS, DG, WHO] • Increased fluid[ACS] • Physical activity[ACS, DG, WHO]
Stroke	• Salt[DG] • Obesity/overweight[DG] • Alcohol[WHO]	• Vegetables, fruits[DG] • Potassium[WHO]
Type 2 diabetes	• Obesity/overweight[DG, WHO] • Saturated fat (meat, dairy products)[WHO] • Sedentary lifestyle[DG, WHO]	• Vegetables, fruits[DG] • Whole grains[WHO] • Legumes[WHO] • Physical activity[DG, WHO] • Dietary fiber[WHO]
Hypertension	• Salt[DG] • Obesity/overweight[DG] • Sedentary lifestyle[DG] • Alcohol[DG]	• Vegetables, fruits[WHO] • Legumes[WHO] • Potassium[DG] • Physical activity[DG]
Obesity	• Sedentary lifestyle[DG, WHO] • Empty-calorie foods, such as sugar-sweetened soft drinks and fruit drinks (high in calories, low in nutrients)[DG, WHO] • Added sugars[DG]	• Physical activity[DG, WHO] • Dietary fiber[DG, WHO] • Whole grains[DG]

Note: Experts are the American Cancer Society (ACS), *Dietary Guidelines for Americans* (DG), and the World Health Organization (WHO).

* Varies by site. See table 3, p. 34, for details.

† See "...But Linked to Heart Disease and Various Cancers," p. 44, for updated information about dairy foods and prostate cancer.

vegetables not only are loaded with nutrients, but they also help push less nutritious foods out of our diets.

Help Fight Heart Disease and Stroke

Several studies have found that both men and women who consume the most fruits and vegetables have the lowest levels of bad cholesterol and a reduced incidence of cardiovascular disease—generally 5 to 30 percent lower than those consuming the smallest amounts.[73] Of course, when people eat more produce, they inevitably eat less of something else, possibly meat or another source of saturated fat and cholesterol. Yet fruits and vegetables have benefits on their own, judging from studies that adjusted for meat intake.[74]

A Finnish study found that middle-aged men who ate the most fruits and vegetables had a 41 percent lower risk of dying from heart disease than those who ate the fewest.[75] Similarly, a U.S. study found a 27 percent lower mortality from cardiovascular disease in adults eating fruits and vegetables three or more times daily compared to those eating them less than once a day.[76] A meta-analysis (see "Meta-Analysis Find Vegetarians Have Less Heart Disease," p. 25) of 14 studies found that each increase in fruit and vegetable intake of about 5 ounces—one generous serving—per day was associated with a 16 percent lower mortality from cardiovascular disease.[77]

One way that fruits and vegetables fight cardiovascular disease is by lowering blood pressure.[78] A 15-year-long study of more than 4,000 young men and women found that people who ate more plant foods, especially fruit, were less likely to develop elevated blood pressure. In a meta-analysis that combined seven long-term studies, each additional serving of fruit was associated with an 11 percent decrease in the risk of stroke. Vegetables had a similar effect.

Play a Role in Cancer Prevention

Fruits and vegetables appear to play a modest role in cancer prevention.[79] Eating more of those foods probably reduces the risk of mouth, esophageal, and stomach cancers.[80] The World Health Organization recommends consuming at least 14 ounces (about four servings) per day of fruits and vegetables to reduce the risk of cancer.[81] The National Cancer Institute's 5 A Day for Better Health Program urges people to eat between five and nine servings of fruits and vegetables a day, depending on sex and age. Experts' conclusions about the effect of what we eat on various kinds of cancer are summarized in table 3. In general, fruits, vegetables, and physical activity are associated with lower risks of certain cancers, while alcohol, a sedentary lifestyle, and red meat and dairy foods appear to increase the risk of certain cancers.

Table 3. **Health experts' advice on diet, physical activity, and cancer**[82]

Cancer site	What increases risk	What decreases risk
Bladder		● Increased fluid intake[ACS]
Breast	● **Overweight or obesity in postmenopausal women**[ACS, WHO] ● Alcohol[ACS, DG, WHO]	● Physical activity[ACS]
Colon, rectum	● **Overweight or obesity**[ACS, WHO] ● **Preserved/processed meat**[WHO] ● Red meat[ACS] ● Alcohol[ACS]	● Physical activity[ACS, DG, WHO]
Endometrium	● **Overweight or obesity**[ACS, WHO]	● Physical activity[ACS]
Esophagus	● **Alcohol**[ACS, DG, WHO] ● **Overweight or obesity**[ACS, WHO]	● **Vegetables, fruits**[ACS, DG, WHO]
Gall bladder	● **Overweight or obesity**[ACS]	
Kidney	● **Overweight or obesity**[ACS, WHO]	
Larynx	● **Alcohol**[ACS, WHO]	● **Vegetables, fruits**[DG]
Liver	● **Alcohol**[ACS, WHO]	
Mouth	● **Alcohol**[ACS, DG, WHO]	● **Vegetables, fruits**[ACS, DG, WHO]
Ovary		● Vegetables, fruits[ACS]
Pancreas	● **Overweight or obesity**[ACS]	● **Vegetables, fruits**[ACS]
Pharynx	● **Alcohol**[ACS, WHO]	● **Vegetables, fruits**[DG]
Prostate	● Dairy products (high-fat*)[ACS] ● High calcium intake mainly through supplements[ACS]	
Stomach		● **Vegetables, fruits**[DG, WHO]

Notes: Experts are the American Cancer Society (ACS), *Dietary Guidelines for Americans* (DG), and the World Health Organization (WHO). Stronger associations are in **boldface**.

* See "...But Linked to Heart Disease and Various Cancers," p. 44, for updated information about dairy foods and prostate cancer.

Walter Willett, chair of the nutrition department at the Harvard School of Public Health, sums up the evidence this way:

> Advice to eat five servings per day of fruits and vegetables…remains sound because a modest reduction in cancer risk is likely, and benefits for cardiovascular disease have become even better established. However, no one should expect substantial reductions in cancer incidence from eating more fruits and vegetables without attention to cigarette smoking, weight control, and regular physical activity.[83]

Help in Weight Loss

With obesity such a major problem in industrialized, and even many developing, nations, scientists have tried to identify the foods that contribute to or prevent weight gain. Intervention studies indicate that substituting fruits and vegetables for foods with higher-calorie densities—such as fatty meats, cheese, and candy—can help with weight loss.[84] Beth Carlton Tohill, of the U.S. Centers for Disease Control and Prevention, notes that "Dietary interventions of low ED [energy-density] diets (low fat and high in fruits and vegetables) led to spontaneous weight loss."[85]

Longer-term observational studies also indicate that eating more produce can fend off weight gain. Four studies involving more than 100,000 adults in all reported an association between higher fruit and vegetable intake and lower weight.[86] Similarly, a survey of more than 420,000 American adults found that people with a normal weight consume

more fruits and vegetables than people who are overweight or obese.[87] (Some studies unfortunately count fried potatoes and fruit juice, which are high in calories, along with "real" fruits and vegetables, obscuring links between the healthiest fruits and vegetables and body weight.[88])

Boost Other Health Benefits

Some studies suggest that diets high in fruits and vegetables are associated with a reduced risk of type 2 diabetes and greater bone density.[89] The World Health Organization notes that such nutrients as vitamin K, potassium, manganese, and boron, all

Super-Star Fruits and Vegetables

It is possible that only specific fruits and vegetables, rather than those entire food groups, reduce cancer risks. For example:

- Tomatoes, possibly because of their carotenoid lycopene, are associated with a reduced risk of prostate cancer.[90]
- Citrus fruits and other sources of the carotenoid beta-cryptoxanthin may reduce the risk of lung cancer.[91]
- Cruciferous vegetables such as broccoli and cauliflower may protect against bladder cancer.[92]

Such findings have led researchers to urge people to focus especially on eating more of certain vegetables and fruits.[93] For instance, the *Dietary Guidelines for Americans*, the U.S. government's authoritative nutrition advice, recommends increased consumption of dark green vegetables, orange vegetables, and legumes. Just eating more French fries and iceberg lettuce won't help.

found in fruits and vegetables, are associated with a decreased risk of bone fracture.[94]

The 2005 Dietary Guidelines Advisory Committee concluded:

> Greater consumption of fruits and vegetables (5–13 servings or 2½–6½ cups per day depending on calorie needs) is associated with a reduced risk of stroke and perhaps other cardiovascular diseases, with a reduced risk of cancers in certain sites (oral cavity and pharynx, larynx, lung, esophagus, stomach, and colon-rectum), and with a reduced risk of type 2 diabetes (vegetables more than fruit). Moreover, increased consumption of fruits and vegetables may be a useful component of programs designed to achieve and sustain weight loss.[95]

Those conclusions led to these key recommendations in the *Dietary Guidelines for Americans*:

> Consume a sufficient amount of fruits and vegetables while staying within energy needs. Two cups of fruit and 2½ cups of vegetables per day are recommended for a reference 2,000-calorie intake, with higher or lower amounts depending on the calorie level.

Choose a variety of fruits and vegetables each day. In particular, select from all five vegetable subgroups (dark green, orange, legumes, starchy vegetables, and other vegetables) several times a week.[96]

Whole Grains

Whole grains are grains that have not been processed to remove the high-fiber bran and germ, which contain much of the protein, vitamins, and minerals. Whole grains are excellent sources of B vitamins, vitamin E, fiber, zinc, iron, other minerals, and a multitude of phytochemicals; these last are naturally occurring chemicals in plants. Many of those substances are largely lost when grain is refined, leaving mostly starch behind. In the United States, four of the B vitamins and iron are added back to "enriched" grains, but that does not fully compensate for the losses.

While the average American eats 11 servings of grains daily, only 1 of those servings is whole grain. Only 7 percent of Americans eat at least three servings a day of whole grains.[97] Instead, virtually all of the grain foods we eat (bread, pasta, cereals, crackers, cookies) are made from white flour, rice is usually white rice, and corn meal is usually degermed.

Decrease the Risk of Cardiovascular Disease

Eating more whole grains appears to reduce the risk of both heart disease and stroke.[98] James W. Anderson, at the Metabolic Research Center at the University of Kentucky, did a meta-analysis of 13 epidemiology studies and concluded that people who ate the most whole grains had a 29 percent lower risk for heart disease than those who ate the least. Even more impressive was the benefit of whole grains in women who never smoked. The Nurses' Health Study found that non-smokers who consumed about three servings of whole grains a day had only half the risk of developing heart disease as women who almost never ate whole grains. In addition, eating more whole grains was associated with a one-third lower risk of ischemic stroke—the kind of stroke that occurs when a blood clot blocks an artery in the brain.

Decrease the Risk of Diabetes

Three large epidemiology studies indicated that whole grains strongly protect against diabetes.[99] The risk of diabetes was about 25 percent lower for people who ate the most whole grains. Another study found that overweight adults who ate 6 to 10 servings of whole grains a day had lower insulin levels than when they ate refined grains.[100] Lower insulin levels can reduce the risk of both diabetes and heart disease.

Clean Out the System

Finally, eating more whole-grain foods could spare millions of people from constipation and other gastrointestinal problems. The fiber in whole grains reduces constipation by increasing fecal bulk, softening stools, and speeding the passage of food through the intestinal tract.[101] In contrast, low-fiber diets lead to hard stools that require a great deal of straining to pass. That straining can lead to increased pressure in the colon and result in diverticular disease (diverticulosis and diverticulitis) and hemorrhoids.[102]

While fiber-rich whole grains were once thought to prevent colon cancer, recent studies indicate that that is unlikely.

Nuts—Protect Against Heart Disease

Several studies strongly suggest that nuts (including peanuts,* which account for two-thirds of all the nuts Americans consume) protect against heart disease.[103] In one study, individuals who ate nuts one to four times weekly had a 22 percent lower risk of heart attack than those eating nuts less than once a week. Eating nuts five or more times per week was associated with a 51 percent lower risk. Those results were consistent in men and women and in younger and older people. In other studies, both walnuts and almonds had a cholesterol-lowering effect when they replaced meat, cheese, or other dairy products.

Nuts' health benefits are likely due, in part, to their monounsaturated and polyunsaturated fatty acid content; those fats lower LDL cholesterol.[104] Other factors may be involved as well, because, as Penny Kris-Etherton and her colleagues at Penn State University found, the effects of nuts on blood cholesterol are greater than predicted on the basis of their fat composition. Other compounds in nuts that may protect against heart disease include dietary fiber, vitamin E, folic acid, copper, magnesium, potassium, arginine, phytochemicals, and plant sterols. The Food and Drug Administration (FDA) has concluded that "Scientific evidence suggests but does not prove

---

*Peanuts technically are legumes, which are discussed below.

that eating 1.5 ounces per day of most nuts as part of a diet low in saturated fat and cholesterol may reduce the risk of heart disease." The main concern about nuts is their high calorie content. It's easy to eat too many nuts, which could lead to weight gain.

Legumes (Beans)—Lower the Risk of Heart Disease

Legumes include dried beans and peas, such as pinto beans, kidney beans, chickpeas, soybeans, and peanuts. (Peanuts account for almost half of all the legumes Americans eat.[105]) Legumes are nutritional powerhouses, and greater consumption of them is strongly associated with a lower risk of heart disease.[106] James W. Anderson and his colleagues at the University of Kentucky and the Veterans Administration Medical Center in Lexington, Kentucky, reviewed 11 clinical trials that examined the effects of legumes (not including soybeans and peanuts) on blood

lipids. They found that total blood cholesterol and LDL cholesterol levels dropped by 6 to 7 percent when 1½ to 5 ounces per day of navy beans, pinto beans, chickpeas, kidney beans, or lentils were included in the usual diet.[107] (In some studies, legumes replaced pasta or other starchy foods, while in others they were just added to the diet.) The Kentucky researchers speculated that legumes' soluble fiber, vegetable protein, folic acid, thiamin, oligosaccharides, and antioxidants may all play a role.

Researchers at the Tulane University School of Public Health and Tropical Medicine found that men and women who ate legumes four or more

Soy Foods: No Health Miracle

Some people claim that soybeans—a dietary staple in China, Japan, and certain other Asian countries—reduce the risk of heart disease, cancer, osteoporosis, and other conditions. However, recent studies indicate that the effect of soy products on LDL (bad) cholesterol levels and blood pressure is trivial.[108] It is not clear whether they protect against cancer.[109] Soy foods also do not appear to benefit postmenopausal women in terms of bone density and cognitive function.[110] Soy products' main health benefit may be that those foods can replace animal products that are high in saturated fat and cholesterol.

times per week had a 22 percent lower risk of heart disease than people who ate legumes less than once a week.[111] The reduced risk of heart disease was not just due to the bean-eaters' eating less meat and poultry. Another study, this involving people in Japan, Sweden, Greece, and Australia, found that every 20-gram increase in daily legume consumption was associated with about a 7 percent lower risk of death.[112]

Beef and Other Red Meat

Beef is a rich source of important nutrients, including protein, vitamin B12 and other B vitamins, iron, and zinc. Heme iron, the form of iron found in meat, fish, and poultry, is easily absorbed and can help maintain iron status and prevent anemia. Zinc is also well absorbed from meat. Unfortunately, those nutrients—which can also be obtained from plant sources, fortified foods, or dietary supplements—are often accompanied by hefty amounts of saturated fat and cholesterol. Nutritionally, that is beef's Achilles' heel.

The Fatty Flaws of Meat and Dairy Foods

Saturated Fat

Animal products account for at least half of the saturated fat Americans eat every day (palm oil also is high in saturated fat). The figure shows the top 10 sources of saturated fat by percentage.[113]

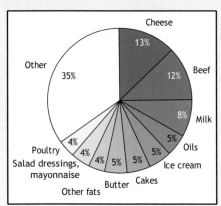

Saturated fat boosts LDL cholesterol in blood, thereby increasing the risk of heart disease.[114] Some studies also link saturated fat to diabetes.[115] In contrast, unsaturated fats from liquid vegetable oils protect against heart disease.

Since even small amounts of saturated fat increase the risk of heart disease, and there is no need for that fat in the diet, the Institute of Medicine of the National Academy of Sciences did not set a "safe" intake level.[116] However, because small amounts of saturated fat occur in everything from corn oil to whole wheat bread, it is impossible and even undesirable for people to try to reduce their intake to zero. The *Dietary Guidelines for Americans* recommends that healthy people consume no more than 10 percent (compared to the current 11 percent) of their calories from saturated fat. People with elevated LDL choles-

Raise the Risks of Heart Attack and Hypertension

Some studies have linked high meat consumption to an increased risk of chronic disease. For example, Frank Sacks and his colleagues at Harvard and the Framingham Heart Study added about 8 ounces of beef to the daily diet of strict vegetarians (vegans) in place of an amount of grains that provided the same number of calories. After four weeks, the subjects' average blood cholesterol level rose 19 percent.[121] Presumably, beef's saturated fat and cholesterol were the culprits. Lean beef likely would have had a smaller effect.

Beef's role in causing heart disease was also indicated in a study of Seventh-day Adventists, as noted earlier. Men who consumed beef three or more times per week had more than twice the risk of a fatal heart attack as men who never ate beef.[122]

Beef consumption also boosts blood pressure.[123] Sacks's clinical study mentioned above found that replacing grains with beef increased the sub-

---

terol are advised to limit their intake to 7 percent of their calories (that is actually good advice for everyone).[117] The best way to cut back is to eat less fatty dairy products and meat.

Cholesterol

Cholesterol occurs only in animal products, including egg yolks, dairy products, shellfish and fish, and meat. The figure shows the top 10 sources of cholesterol by percentage. Cholesterol is in both the lean and fatty parts of meat, so choosing lean meat helps lower saturated fat, but not cholesterol, intake. Our bodies produce all the cholesterol they need.

Dietary cholesterol increases LDL cholesterol levels in blood and the risk of heart disease.[118] The average cholesterol intake for middle-aged (19-50 years old) men is around 350 milligrams and for women 210 milligrams.[119] While official recom-

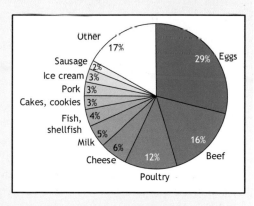

mendations[120] are to limit cholesterol to 300 milligrams daily—200 milligrams or less for people with elevated LDL cholesterol—the less cholesterol consumed, the better. However, small amounts—from poached salmon, skinless chicken, or even an occasional egg yolk—are not a problem.

jects' systolic blood pressure by 3 percent in just four weeks. When thousands of women were monitored for several years in the Nurses' Health Study, eating more beef and processed meats was associated with a higher systolic blood pressure. Furthermore, a study of several thousand young adults found that people who ate more beef and pork were likelier to develop elevated blood pressure.

Raise the Risk of Cancer

According to the American Cancer Society, red meat may increase the risk of cancer of the colon and rectum.[124] Similarly, the World Health Organization advises that high intakes of red and processed meats "probably" increase the risk of those cancers.

A major study by the American Cancer Society examined more than 148,000 adults who had provided dietary information 9 and 19 years earlier.[125] People who ate the most beef and pork at both points in time had the highest risk of rectal cancer. Those who ate the most processed meat—such as ham and bacon—also had a higher risk of cancer in the part of the large intestine closest to the rectum.

In other studies, Seventh-day Adventist men and women who ate red meat one or more times a week had almost twice the risk of colon cancer as those who never ate red meat.[126] The EPIC study, which is tracking diet and disease in half a million Europeans, found that eating more red meat

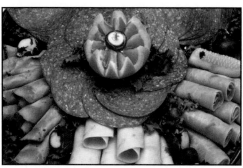

Consumption of red meat—especially processed meats—increases the risk of colon cancer and possibly cancer of the pancreas.

and processed meat increases the risk of colorectal cancer.[127] Eating more than about 7 ounces per day of those meats was associated with a one-third increase in colon cancer. Processed meat appears to be more harmful than red meat.

A meta-analysis of almost two dozen studies indicated a 35 percent increased risk of colon cancer in people who ate red meat and a 31 percent increase in people who ate processed meat compared to those who ate little or no meat.[128] Every 3½-ounce-per-day increase in consumption of red meat was associated with about a 15 percent increased risk of colon cancer, while every 1-ounce increase in daily consumption of processed meat was associated with almost a 49 percent higher risk.[129]

New research suggests that red meat—especially processed red meat—also increases the risk of pancreatic cancer.[130] Whether it's the meat itself or contaminants or additives introduced into the meat during processing is not yet known.

Increase the Risk of Diabetes

The Harvard School of Public Health's study of over 40,000 male health professionals found that processed meat appears to increase the risk of diabetes. Harvard's parallel Nurses' Health Study, involving almost 70,000 women, found that diabetes was linked strongly to consumption of bacon, and less strongly to hot dogs, other processed meats, and red meat.[131]

Poultry

Fortunately, considering how much of it we are eating these days, poultry does not appear to contribute directly to chronic disease. Indeed, poultry is usually lower in fat and saturated fat than red meat, so it is much healthier to eat in terms of heart disease. On the other hand, much of the chicken Americans eat has been deep-fried by restaurants in partially hydrogenated oil, the major source of heart-damaging trans fat, and heavily salted. That's plain old junk food.

Dairy Products

Healthy, to a Point...

Dairy products are excellent sources of calcium, and fluid milk is an excellent source of vitamin D. Those nutrients, along with potassium, are needed to build and maintain bones at all ages. A number of studies have found that people who consume more dairy products have stronger, denser bones, and thus a lower risk of developing osteoporosis ("brittle bone" disease) or of fracturing a bone.[132] The protective effect of dairy products is strongest in younger women and less significant in women over 50. (There is lim-

Replacing Elsie Healthfully

Dairy products are Americans' biggest sources of calcium, vitamin D, and potassium. But people who choose to eat little or no dairy foods can get calcium from green leafy vegetables, tortillas processed with lime, fortified foods, and supplements. They can get vitamin D from fortified foods (soymilk, breakfast cereals), supplements, or exposure to sunlight. Potassium is widely distributed in fruits and vegetables. That said, dairy foods may contain unique compounds that people who eschew those foods would miss out on.[133]

ited information on the benefits to men's bone health of eating more dairy products.[134])

Dairy products also appear to help the body regulate blood pressure, thereby reducing the incidence of hypertension.[135] Adding three servings per day of low-fat dairy products to a healthy diet (low in saturated fat and total fat and high in fruits and vegetables) reduces blood pressure. Consistent with that, dairy products have been associated with a reduced risk of stroke and metabolic syndrome (a group of symptoms including obesity and insulin resistance that increases the risk for heart disease and type 2 diabetes).

...But Linked to Heart Disease and Various Cancers

Whole milk and many cheeses are major sources of saturated fat and cholesterol, which cause heart disease. Milk and cheese account for 21 percent of the saturated fat and 11 percent of the cholesterol in the American diet.[136] Cheese is now the single greatest source of saturated fat. Considering that whole milk has been a major (though declining) source of saturated fat in the American diet, it is no surprise that studies have correlated higher consumption with heart attacks. In the Nurses' Health Study, women who drank two or more glasses of whole milk a day had a two-thirds greater risk of fatal and nonfatal heart attacks than women who drank less than one glass a week.[137]

People could (and should) switch to fat-free dairy products. However, from a public health perspective, that doesn't lower the overall risk of heart disease, because the milkfat ends up in cheaper butter or in cream, premium ice cream, and other high-fat foods. Suggestions for ways to lower the fat or saturated fat content of cow's milk are discussed in "Reduce the Fat Content of Milk," p. 154.

Dairy products also have been associated with increased or decreased risks of certain cancers.[138] Eight of 11 studies that monitored large groups of men over a number of years linked dairy foods to an increased risk of prostate cancer. Researchers at the Harvard School of Public Health stated that the association between dairy products and prostate cancer is one of the more consistent dietary predictors for prostate cancer. According to one estimate, eating three servings per day of dairy products is associated with a 9 percent greater risk of prostate cancer—or 20,000 more cases per year.[139] (Note, however, that most men actually consume about half that many servings.[140])

Just what it is in dairy products that might promote prostate cancer is not known. The fat does not appear to be the problem, because several studies linked skim and low-fat milk to prostate cancer.[141] Several studies have suggested that calcium is the culprit, although others dispute this.[142]

While dairy foods might promote prostate cancer, they also might reduce the risk of other cancers. Some studies have found a modestly lower risk of colorectal cancer in people who consumed more milk.[143] (Cheese and yogurt did not appear to protect against cancer.) And dairy products may reduce the risk of breast cancer in premenopausal women, but the evidence is inconsistent.[144] But while we await more research, we need to eat. Considering dairy products' pluses and minuses, it makes sense to consume two or three servings of them a day, but not go overboard with five or six.

Eggs

Keep the Whites, Toss the Yolks...

The main health concern with eggs is their effect on heart disease. The problem is not whole eggs, but the yolks. Egg yolks supply close to 30 percent of the 270 milligrams of cholesterol in the average adult's daily diet.[145] While 270 milligrams is within the "less than 300 mg per day" guideline, the 2005 Dietary Guidelines Advisory Committee states that "cholesterol intake should be kept as low as possible, within a nutritionally adequate diet."[146] Dietary cholesterol increases LDL cholesterol in blood, which, in turn, increases the risk of heart disease.[147] Egg whites, in contrast, are rich in protein and free of cholesterol.

...To Reduce the Risk of Heart Disease

The high cholesterol content of egg yolks implies that egg-rich diets would increase the risk of heart disease—and studies of populations indicate that eggs do exactly that. For example, the Oxford Vegetarian Study found that eating eggs more frequently was associated with a substantial increase in the risk of death from heart disease.[148] Dutch researchers conducted a meta-analysis of 17 well-controlled studies on the effect of dietary cholesterol from eggs on the ratio of total blood cholesterol to HDL (good) choles-terol.[149] Many experts consider that ratio to be one of the best indicators of heart-disease risk, with higher ratios indicating greater risks. In all but one of the studies examined, the researchers found that increased egg consumption was associated with higher ratios. Finally, men and women with diabetes have an increased risk of heart disease as their egg con-sumption increases.[150] The bottom line is that we should eat fewer egg yolks.

Fish

Decreases Risk of Heart Disease and Cancer

Fish is generally quite healthful, notwithstanding several concerns dis-cussed below. Eating fish reduces the risk of heart disease. A meta-analy-sis of studies involving a total of more than 200,000 people found that those who ate fish at least once a week had a 15 percent lower risk of dying from coronary heart disease than those who ate fish less than once a month. Peo-ple who ate fish five or more times per week had almost a 40 percent lower risk.[151] Of course, frying fish in partially hydrogenated oil—as restaurants often do—turns a dietary plus into a minus.

The health benefits of fish probably come from a favorable mix of fatty acids, including low levels of saturated fat and high levels (in some species) of two omega-3 fatty acids: eicosap-entaenoic acid (EPA) and docosahex-aenoic acid (DHA). Those omega-3s are thought to prevent heart attacks and strokes.[152] The World Health Organization, American Heart Asso-ciation, and *2005 Dietary Guidelines for Americans* all recommend eating at least two servings of fish per week.[153]

Eating fish may also protect against cancer. The large EPIC study in Europe found that people who ate more than 2.8 ounces of fish per day had a one-third lower risk of colorec-tal cancer than those eating little or no fish (under 0.3 ounces).[154] Further-more, several studies indicate that fish may reduce the risk of prostate cancer.[155]

Not Enough Fish

While health experts are encourag-ing people to eat more fish, over-fishing is driving some species to the brink of extinction. Populations of Pacific cod, Atlantic sturgeon, shark, monkfish, numerous variet-ies of rockfish, and others are all in trouble. Even aquaculture is a problem, because some farmed fish, such as salmon, are fed meal made from small ocean-dwelling fish that would otherwise provide food for diverse wild species. Before head-ing for the seafood counter, visit the Monterey Bay Aquarium's Seafood Watch (www.mbayaq.org/).

Some Seafood Contains Dangerous Contaminants

Not everything about fish is salubrious. Contamination of certain species of fish by mercury, polychlorinated biphenyls (PCBs), and dioxins detracts from fish's healthfulness, at least for pregnant and nursing women, infants, and young children. Those pollutants—in fish and other animal products—are discussed later in this chapter (see "What in Animal Foods Harms Us," p. 52). In addition, natural toxins—ciguatoxin and scombrotoxin—in finfish and potentially deadly *Vibrio* bacteria in Gulf of Mexico shellfish cause food poisoning.

What Actually Nourishes Us

A variety of well-known substances in foods contribute to their healthfulness: fiber, antioxidants, folate, and potassium, to name a few. In addition, plants contain thousands of other phytochemicals that may have health benefits. Some of the substances, such as potassium, that are found in plants also occur in animal foods; others, such as fiber and vitamin C, occur only in, or are more abundant in, plants.

Dietary Fiber

All minimally processed plant-based foods contain fiber. Highly processed plant-based foods, such as white flour, sugar, and vegetable oil provide little or no fiber. Animal products—meat, dairy, eggs, and seafood—provide no fiber at all.

Fiber actually encompasses a multitude of different substances. These are typically divided into two broad groups:

● Soluble (or viscous) fiber, commonly found in fruits, oats, barley, and dried beans, dissolves in water and can slow the rate at which food leaves the stomach, which may help with weight control as well as reduce blood glucose levels.[156] Soluble fiber also interferes with the absorption of dietary cholesterol and reduces LDL cholesterol in blood.[157]

- Insoluble fiber, which occurs in whole grains, nuts, and some fruits and vegetables, does not dissolve in water. Cellulose, some hemicelluloses, and lignins are the most common insoluble fibers. Insoluble fiber increases stool bulk, alleviates constipation, and reduces the risk of diverticular disease.[158]

What Fiber Does *Not* Do

Fiber is not a panacea. Researchers at the National Cancer Institute and elsewhere long thought that dietary fiber helped prevent colon cancer.[159] However, several important epidemiology studies and three intervention trials did not find a benefit.[160] Fiber also does not appear to prevent pre- or postmenopausal breast cancer.[161]

Fiber—especially soluble fiber and fiber from grain products—has been consistently linked to a lower risk of heart disease.[162] A long-term Harvard study of male health professionals found that men who ate an average of 29 grams of total fiber per day had half the risk of fatal heart disease as those who ate half as much fiber.[163] A subsequent study conducted by Tulane University scientists found that men and women who consumed about 6 grams of soluble fiber per day had a 24 percent lower risk of dying from heart disease and a 12 percent lower mortality from all causes compared to those who consumed about 1 gram per day.[164] Cereal fiber was more closely associated with the reduced risk than was fiber from fruits and vegetables. In women, eating 5 grams per day more cereal fiber—equivalent to two or three slices of whole wheat bread—was associated with a 37 percent lower risk of heart attack and stroke.[165]

A meta-analysis found that each 10-gram increase in dietary fiber was associated with a 14 percent lower risk of all coronary events and a 27 percent lower risk of death from heart disease. Fiber from cereal and fruit appeared to provide the most benefit, while fiber from vegetables had little effect.[166] The results were similar for men and women.

At a time when millions of people are seeking cures for obesity, it is important to note that people who eat the most fiber tend to weigh less.[167] Dietary fiber helps control weight in several ways:

- Fiber-rich foods have to be chewed more, which slows eating speed.
- Fiber-rich foods take up a relatively large volume in the stomach, making people feel full sooner.[168]
- Soluble fiber slows stomach emptying, which keeps people feeling full for a longer time.[169]

The World Health Organization and others have identified diets high in dietary fiber—that is, rich in whole grains, fruits, vegetables, and beans—as an important means of preventing obesity.[170]

One final virtue of fiber—and a most valuable one—is that it acts as a laxative, leading to softer, bulkier stools. Fiber from wheat bran has the greatest effect, followed closely by fiber from fruits and vegetables.[171]

The Institute of Medicine, a unit of the National Academy of Sciences, recommends that middle-aged (19–50 years old) men consume 38 grams of fiber per day and women 25 grams per day.[172] Currently, the average man and woman consume only half that much. In contrast, American and British vegetarians consume much more: Lacto-ovo vegetarians average 23 grams per day, while vegans average 35 grams per day.[173]

Folate

Folate is a B vitamin found in green vegetables, orange juice, fortified grains, and dried beans. Among other things, this important vitamin helps the body make new proteins, DNA, and red and white blood cells. Consuming too little folate during early pregnancy increases the risk of neural tube defect, a serious birth defect in which the neural tube fails to encase the spinal cord. Folate also may reduce the risk of colon cancer, but more research is needed.[174]

Since 1998, the FDA has required that white flour for bread and pasta, white rice, and breakfast cereals made with refined flours be fortified with folic acid. Previously, adults consumed only about two-thirds of the recommended amount of folate.[175] Fortification has almost doubled Americans' folate intake.[176] Happily, the incidence of spina bifida (one type of neural tube defect) has declined by 20 percent.[177]

Because plant-based foods are rich in folate, vegetarians tend to consume more of the vitamin than non-vegetarians.[178] In 1994–96 (before white flour and white rice were fortified), the average American consumed 262 micrograms of folate per day.[179] The average vegetarian likely consumed at least half again more.[180] Post-fortification comparisons have not been conducted. Despite a higher level of folate, white flour is poorer in many other nutrients and dietary fiber than whole wheat flour. People would be better off eating foods made with whole-grain flour, plus a multivitamin supplement.

Potassium

The mineral potassium is abundant in fruits, vegetables, and beans, as well as in milk and seafood. The median potassium intake of U.S. adults is about 3 grams per day for men and just over 2 grams for women. That is well below the "adequate" level of 4.7 grams per day, generally because of our limited consumption of fruits and vegetables. Some of the richest food sources of potassium are spinach, cantaloupe, almonds, Brussels sprouts, and bananas,[181] but the biggest sources in the American diet are milk, potatoes, coffee, and beef.[182]

Potassium plays an important role in regulating blood pressure.[183] A higher intake of potassium is associated with a lower blood pressure, and increasing potassium can reduce blood pressure in people with or without hypertension. Higher potassium intakes, judging from several studies, lower the risk of stroke.

Consuming more potassium has been associated with greater bone density and less age-related decline in bone density.[184] In addition, a higher potassium intake may well reduce the risk of kidney stones.[185]

Unsaturated Oils

Most fats and oils in plants, including soy, corn, canola, safflower, olive, and sunflower oils, contain beneficial mono- and polyunsaturated fatty acids. Those unsaturated fats lower the bad cholesterol in our blood.[186] (In contrast, of course, animal fats are relatively high in saturated fat and low in unsaturated fatty acids and raise the bad cholesterol.) Based on a study of more than 80,000 women, Harvard researchers estimated that substituting

Canola plants have beautiful flowers, as well as seeds that are rich in healthy monounsaturated oil.

unsaturated fat for about one-third of the saturated fat in a typical diet would reduce the risk of heart disease by a hefty 42 percent.[187] Indeed, Americans are consuming three times as much salad and cooking oils as they were 40 years ago, a dietary change that almost certainly has prevented thousands of fatal heart attacks every year (see "The Cardiovascular Benefit of Eating Less Meat and Dairy," p. 20).[188]

Omega-3 Fatty Acids

Omega-3 fatty acids are a family of polyunsaturated fatty acids that occur in fatty fish, some vegetable oils, soy products, walnuts, and certain other foods. As noted above, the omega-3s probably contribute to the association between eating fish and a lower risk of cardiovascular disease. Plants contain not the EPA or DHA omega-3s that occur in fish, but another omega-3, alpha-linolenic acid. Unfortunately, the body converts only a small fraction of that to EPA.[189] (The body can then convert a small fraction of the EPA to DHA.) It is unclear whether the alpha-linolenic acid reduces the risk of heart disease.

The Institute of Medicine recommends that men consume 1.6 grams of alpha-linolenic acid daily and that women consume 1.1 grams. Anyone who doesn't eat much fish should consume adequate alpha-linolenic acid from flaxseed, flaxseed oil, canola oil, tofu, soybeans, soybean oil, and walnuts and should consider taking a fish-oil or DHA supplement.[190]

Antioxidants

Antioxidants include such nutrients as vitamin C, beta-carotene, vitamin E, and selenium. Fruits and vegetables are especially rich sources of many antioxidants. Researchers have long hypothesized that antioxidants help protect against harmful oxidizing agents, which can damage body proteins, DNA, and fats. While researchers have speculated that antioxidants contribute to the ability of fruits and vegetables to reduce the risk of chronic disease, intervention studies with vitamin C, vitamin E, and beta-carotene have not found any benefits.[191] In fact, smokers who took beta-carotene supplements actually had a *greater* risk of lung cancer.[192]

Fruits and vegetables contain antioxidants that may provide health benefits.

Intervention studies with selenium have been more promising. Several studies found that the mineral lowers the risk of prostate cancer and possibly other cancers, especially in people with low blood levels of selenium.[193]

Antioxidants probably are best acquired from whole foods rather than from dietary supplements.[194] It may turn out that it is not antioxidants but other constituents of plants that are the truly beneficial substances.

Phytochemicals

Phytochemicals can be loosely defined as any chemicals that are naturally present in plants. Scores of different phytochemicals have been identified in fruits, vegetables, legumes, whole grains, and nuts. General categories of phytochemicals include carotenoids, flavonoids, isoflavones, lignans (not to be confused with lignins, which are plant fibers), and phytosterols. Many of them have no effect at all on health, but initial studies suggest that some reduce the risk of heart disease, cancer, stroke, cataracts, and other diseases.[195] While researchers work out the details, consumers should just eat plenty of a wide variety of fruits, vegetables, whole grains, and nuts and not bother taking supplements that are costly and contain just a few cheap and convenient phytochemicals. Much exciting new research is exploring the exact role of phytochemicals in disease prevention.

What in Animal Foods Harms Us

Although some animal products are rich sources of protein, calcium, iron, zinc, and other essential nutrients, many also are rich sources of potentially harmful components, including saturated fat and cholesterol (see "The Fatty Flaws of Meat and Dairy Foods," p. 40). In addition, chemical by-products of cooking and environmental toxins—such as heterocyclic amines (HCAs), polycyclic aromatic hydrocarbons (PAHs), PCBs, and pesticides—often occur in fatty animal products and may be harmful.

Heterocyclic Amines and Polycyclic Aromatic Hydrocarbons

HCAs form when meat, poultry, fish, and eggs are cooked at high temperatures, especially by grilling or frying. HCAs are potent mutagens (agents that cause genetic mutations) that cause cancer in animals.[196]

Another group of chemicals, polycyclic aromatic hydrocarbons, form when fat from meat, poultry, and fish drips onto hot coals or a flame. PAHs are created and then rise with the smoke, contaminating the food.[197] The nutrition-oriented World Cancer Research Foundation identified grilled and barbecued meats and fish as possible causes of stomach and colon cancer.[198] The National Toxicology Program of the U.S. Department of Health and Human Services says that PAHs and four HCAs are "reasonably anticipated to be human carcinogens."[199] As with other carcinogens, the more of those substances that are consumed, the greater the risk, but the risk from using the backyard grill a few times over the summer is trivial.

Environmental Contaminants

Environmental contaminants, including pesticides (see "Risks from Pesticides," next page), industrial chemicals, and various pollutants, often accumulate in animal fat. That is why meat, full-fat dairy products, and fatty fish tend to be the major sources of those contaminants. Fat-soluble contaminants persist for many years in human (or other animals') fatty tissue and occur in breast milk. Some of the contaminants cause cancer in experimental animals and appear to cause behavioral abnormalities in humans.

Risks from Pesticides

Pesticides are widely used to control insects, weeds, and fungi on cropland and crops. Of the 511 million pounds of pesticides used in 2001, 181 million pounds were used on crops for livestock.[200] The vast majority—167 million pounds—was used for feed grains, with the remainder for hay and pasture.

It is difficult to determine the health effects of pesticides on consumers, because the levels of pesticide residues in food and water are minuscule and the effects may be rare or subtle. No one expects there to be a trail of sick people leading from the dinner table to the hospital. Rather, the concern is that long-term exposure to low levels of numerous pesticides may cause diseases ranging from autism to cancer or impair the immune system.[201]

Animal Studies: Raising Concerns

Many pesticides, including alachlor, acetochlor, and atrazine—herbicides widely applied to animal feed crops—have caused tumors in laboratory animals, including stomach tumors in male rats and stomach, lung, and mammary tumors in female rats.[202] When a chemical causes tumors in animals, it is presumed to pose a cancer threat to humans. Other pesticides, such as carbaryl and methyl-phenoxyacetic acid, when tested at high doses suppressed the immune systems of lab animals and may cause autoimmune disorders; they also damaged the spleens, livers, kidneys, and nervous systems of the animals.[203] While those results are intriguing, if not downright scary, epidemiology studies can help clarify whether the chemicals actually harm humans.

Farmers: The Canary in the Coal Mine

Farmers, not by choice, serve as important indicators of health risks from pesticides. And some of the studies on farmers who regularly apply pesticides suggest significant risks. While farmers have lower overall rates of cancer than the general population, due to factors such as less smoking, they have higher rates of several cancers (see figure).[204]

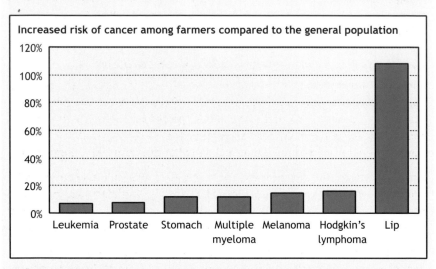

Increased risk of cancer among farmers compared to the general population

Note: Results are based on surveys of farmers in the United States and abroad.[205] Increases are statistically significant.

Many factors contribute to the higher cancer rates among farmers. For instance, the higher rate of melanoma, a serious form of skin cancer, may be from working long hours in direct sunlight. Pesticide exposure also appears to be a significant factor.

● California researchers Paul K. Mills and Richard Yang concluded that the higher risk of prostate cancer in Hispanic farmworkers was related to their exposure to certain herbicides.[206]

● The National Cancer Institute's Agricultural Health Study, which involves nearly 90,000 participants in Iowa and North Carolina, associated an increased risk of prostate cancer with six different pesticides.[207]

● Nebraska farmers who applied 2,4-dichlorophenoxy acetic acid more than 20 days a year were three times as likely to develop non-Hodgkin's lymphoma as farmers not exposed to the pesticide.[208] That herbicide is often applied to almost every major grain and roughage crop fed to livestock.[209]

That said, some studies did not find clear links between two widely used herbicides—atrazine and glyphosate—and cancer.[210]

Organophosphate pesticides are highly neurotoxic, causing weakness and even paralysis.[211] They make up nearly half of all insecticides (some also serve as herbicides and fungicides) used in the United States, with some 5 million pounds annually applied to corn, hay, soybean, wheat, pasture, and other crops.[212] Heavy exposure of farmworkers (and others) to organophosphates has been linked to memory loss, confusion, limb paralysis, and behavioral abnormalities, as well as paralysis of the lungs (which causes death by suffocation).[213] One study linked the now-banned soil fumigant 1,2-dibromo-3-chloropropane to reduced or absent sperm production in farmworkers.[214]

Health Effects in Consumers: The Big Question Mark

Consumers are exposed to far lower levels of pesticides than farmers and pesticide applicators, so the risks to them are far smaller. However, children are particularly vulnerable, because they metabolize certain pesticides differently from adults and they consume higher concentrations of pesticides relative to their weight.[215] Subtle impairment of IQ or behavior would be of great import, but undetectable.

While most consumers are especially concerned about pesticide residues on fruits and vegetables, which are directly sprayed, fat-soluble pesticides in animal products actually pose the bigger risk. That's because livestock are fed large amounts of pesticide-tainted feed grains and accumulate pesticide residues in their fat.

Agricultural pesticides may well be causing the same problems in consumers as they cause in lab animals and farmworkers, albeit at a much lower frequency and severity. Much of the concern focuses on weakening the immune system, which might lead to higher rates of infectious diseases and cancer. That is especially true for people, such as Inuit children, whose diets contain high levels of pesticides and toxic chemicals.[216] However, actually proving that the average consumer is harmed may be impossible. Consumers could reduce their exposure to pesticides by purchasing foods produced on organic farms or farms that use integrated pest management and by eating less meat, poultry, dairy, and egg products overall.

Polychlorinated biphenyls are highly toxic industrial chemicals that are "reasonably anticipated to be human carcinogens."[217] In addition, PCBs endanger fetuses and young children because they can affect the developing brain. The contaminant concentrates in animal fat and enters our diets and bodies through fish, cheese, eggs, and other foods. A 2003 report by the nonprofit Environmental Working Group found that samples of farmed salmon from the East and West Coasts of the United States contained three times as much PCBs as typical commercial seafood and about four times more PCBs than beef.[218]

PCBs in farmed salmon are a problem for children and pregnant women, but less so for others. For the average adult the cardiovascular benefit from omega-3 fatty acids in salmon far outweighs the cancer risk from PCBs. The Center for Science in the Public Interest estimates that if 100,000 people ate one serving of farmed salmon per week, one person would develop cancer, but 1,500 people would be spared death from cardiac arrest.[219]

Another fat-soluble industrial contaminant that lurks in food—and human blood and breast milk—is polybrominated diphenyl ethers (PBDEs). According to Arnold Schechter and his colleagues at the University of Texas Health Science Center in Dallas, "food is a major route of intake for PBDEs," with fish, cheese, butter, and poultry being the most contaminated.[220] These chemicals, which are used as flame retardants in everything from furniture foam to plastics in personal computers, are chemically and toxicologically similar to PCBs. The Environmental Protection Agency has reported that PBDEs cause liver and thyroid toxicity, as well as neurodevelopmental problems.[221] The agency banned the most toxic types of PBDEs in 2005, but residues of these chemicals will persist in the environment—and food supply—for many years to come.

Mercury

Mercury is a toxic metal spewed into the air by coal-burning power plants and carried around the globe. It accumulates in the tissues of fish, especially large predatory fish. Like PCBs, mercury is especially toxic to fetuses and young children and can cause irreversible neurological damage. That's why the Environmental Protection Agency and the FDA recommend that women who are or may become pregnant, nursing women, and young children completely avoid shark, swordfish, and king mackerel. Other fish and shellfish should be limited to 12 ounces (6 ounces for albacore tuna) per week.[222]

What It All Means

Over the past half-century, hundreds of studies—animal, clinical, epidemiological, and intervention—have examined the effect of diet on health from every conceivable angle. They provide strong, consistent evidence that diets rich in animal foods (except fish)—especially fatty meat and dairy products—and poor in healthy plant-based foods contribute to hypertension, stroke, heart disease, cancer, obesity, and diabetes. That rich body of research has led the world's leading health experts to emphasize the benefits of plant-based diets. The World Health Organization, the U.S. government's *2005 Dietary Guidelines for Americans,* the American Academy of Pediatrics, the American Heart Association, the American Cancer Society, and many other authoritative health agencies all recommend that people eat more fruits, vegetables, and whole grains and modest amounts of non-fried seafood and poultry, low-fat dairy products, and lean meat.[223] Most health experts also strongly recommend that people cut back on salt, refined sugars, and partially hydrogenated oils. Eating more of the healthy foods provides essential nutrients *and* squeezes less-healthy foods off the plate. In "Changing Your Own Diet" (p. 143), we provide more specific advice on what precisely a healthy diet should include, along with a scorecard for evaluating your diet.

Meatless burgers are being made of everything from chickpeas (above) to mushrooms and oats to soybeans.

Less Foodborne Illness

A potent case of food poisoning, with its nausea, vomiting, and I-think-I'm-going-to-die misery, is unforgettable. Some of the most common causes of food poisoning are the bacteria and viruses carried by farm animals and that are abundant in their manure. Many common germs live harmlessly in animals but can make people deathly ill. Those pathogens can jump from animals to people through tainted food, air, soil, water, or direct contact between people and livestock.

Although diet-related chronic diseases, such as heart disease and various kinds of cancer, kill many more people than food poisoning, the sudden onset of food poisoning and the fact that it can be traced to particular foods add urgency to efforts to control it.

- More than 1,000 Americans die each year from foodborne illnesses linked to meat, poultry, dairy, and egg products.

- The annual medical and related costs of foodborne illnesses in the United States are at least $7 billion.

- Fruits and vegetables are a major cause of food poisoning thanks, in part, to contamination from livestock manure.

- Raising large numbers of poultry and pigs increases the risk of deadly flu epidemics.

The Scope and Costs of Foodborne Illness

Foodborne illnesses are caused by such well-known bacteria as *Campylobacter jejuni,* the deadly O157:H7 strain of *Escherichia coli* (*E. coli*), and several types of *Salmonella,* as well as by such little-known germs as Norwalk-like viruses. The federal Centers for Disease Control and Prevention (CDC) estimates that pathogens in food cause about 76 million illnesses, 325,000 hospitalizations, and 5,200 deaths each year (see table 1).[1] Norwalk-like viruses, which cause gastrointestinal distress, are the most common source of illnesses whose causes have been identified. Typically, they are transferred to food by poor sanitary practices during preparation. Although bacteria cause fewer illnesses than viruses, they are more likely to be fatal. In fact, listeriosis, caused by *Listeria monocytogenes,* is fatal in 20 percent of the people it infects. Germs, such as *E. coli* and *Salmonella,* associated with food animals accounted for at least 1,100 of the deaths (and probably many more in the "unknown" category).

Table 1. **Major causes and costs of foodborne illnesses and deaths in the United States (annual estimates)**[2]

Pathogen		Illnesses	Deaths	Cost (medical, lost productivity, premature death)
Bacteria	*Campylobacter*	2,000,000	100	$1.2 billion
	Salmonella	1,300,000	550	$2.4 billion
	Clostridium perfringens	249,000	7	Not available
	Staphylococcus	185,000	2	$1.2 billion
	E. coli*	94,000	80	$1.0 billion
	Listeria	2,500	500	$2.3 billion
Parasites	*Cryptosporidium parvum*	300,000	7	Not available
	Giardia[†]	200,000	1	$0.5 billion
	Toxoplasma gondii	113,000	380	Not available
Viruses	Norwalk-like viruses	9,200,000	120	Not available
Unknown		62,000,000	3,200	Not available
	Total[‡]	76,300,000	5,207	$6.9 billion

* The estimate covers only O157:H7 and other shiga toxin-producing strains of *E. coli.* Other strains of *E. coli* cause additional illnesses.

† Although cattle carry *Giardia,* it is unclear whether they carry strains that can infect humans. The deaths may be due to *Giardia* from wildlife or other sources.

‡ Figures do not sum to totals because data are limited to those pathogens causing in excess of 100,000 illnesses or 80 deaths. See source for complete listings. Moreover, $6.9 billion probably is an underestimate because the costs of many major foodborne illnesses never have been calculated; this total covers only about 9 percent of the estimated 76 million foodborne illnesses suffered each year.

The causes of food poisonings are rarely tracked down, because it is not worth the effort and cost when only single individuals are affected. Instead, public health experts focus on outbreaks affecting dozens or hundreds of people. The Center for Science in the Public Interest (CSPI) has compiled a database of 3,810 outbreaks caused by germs (plus another 700 caused by toxins in fish) and for which the contaminated food was identified.[3] Though the database covers only a small percentage of foodborne illnesses, it indicates which foods pose the greatest risks.

Red meat and poultry—including luncheon meats—caused more than 1,200 of the outbreaks in CSPI's database (see table 2). Americans eat far less seafood than meat and poultry, yet seafood was linked to more than 300 of the outbreaks. Fruits and vegetables, normally thought of as being perfectly safe, caused over 500 of the identified outbreaks. However, about one-third of those outbreaks actually were caused by germs normally associated with animal manure, as were two-thirds in the "other" category. Dairy was the safest major category in the database, causing about 150 of the identified outbreaks, but the largest outbreaks on record were caused by dairy foods.[4] In 1985, milk contaminated with *Salmonella* sickened over 16,000 people in the Chicago area and killed 2. In 1994, 224,000 people around the country were sickened by ice cream made from ingredients contaminated with *Salmonella* that was in dirty tanker trucks. All told, 58 percent of the outbreaks were associated with animal products or germs normally associated with livestock.

Considering how much of our food is contaminated, it is remarkable that foodborne illnesses do not strike more people. In 2002, *Consumer Reports*

Table 2. **Sources of foodborne illness outbreaks in the United States linked to microbial hazards, 1990–2003**[5]

Food	Outbreaks		People sickened	
	Number	Percent	Number	Percent
Meat, poultry, luncheon meats	1,221	31	38,284	28
Seafood*	306	8	6,609	5
Vegetables and fruits	529	14	28,108	20
Eggs	329	9	10,849	8
Dairy	151	4	5,145	4
Other (sandwiches, pasta, salads, ethnic foods, etc.)	1,274	33	49,667	36
Total[†]	3,810	100	138,662	100

* Includes only microbial-linked outbreaks, not those due to scombroid or ciguatera toxins in fish.

[†] Percentages do not total 100 because of rounding.

magazine found that 1 percent of ground beef samples bought at grocery stores had significant levels of fecal contamination and 4 percent were on the brink of spoilage.[6] The magazine's tests of almost 500 fresh chickens from 25 cities found that 42 percent were contaminated with *Campylobacter* and 12 percent with *Salmonella*.[7] Overall, 49 percent of the chickens were

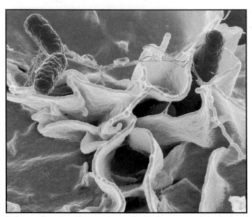

Salmonella *is the main culprit in egg-related food poisonings and is a common contaminant in chicken and meat. Shown here (pink) growing on cultured cells.*

contaminated with one or both bacteria. Adding to the risk, 90 percent of the *Campylobacter* and one-third of the *Salmonella* were resistant to at least one antibiotic. The U.S. Department of Agriculture (USDA) found an even bigger problem: 90 percent of birds tested positive for *Campylobacter*.[8] Government data also show that about 2.3 million eggs are contaminated with *Salmonella* each year.[9] Although thorough cooking kills the *Campylobacter* and *Salmonella* in infected meat, poultry, and eggs, the contaminated raw foods may infect consumers who touch them or eat them undercooked.

Foodborne illnesses typically occur shortly after tainted foods are eaten and, while causing real misery, are short-lived. But they sometimes have long-term consequences. Guillain-Barré Syndrome, an autoimmune disorder caused by *Campylobacter* infection, is one such lingering result. Reiter's Syndrome is a type of arthritis caused by *Salmonella*. Even more disturbing than those relatively rare events is what a study at the Statens Serum Institute in Denmark found. These scientists tracked 49,000 people who had suffered gastrointestinal infections and compared them to individuals who had not. The findings? People who had had food poisoning were more than three times as likely to die in the following year.[10] In other words, individuals who contract foodborne illnesses are either already in poor health—or foodborne illnesses may be much more harmful than anyone thought.

Our Food System Increases Certain Food-Safety Risks

Food poisoning has afflicted humans since time immemorial and was considered an inevitable part of life. Health officials and industry have improved the safety of the food supply through the use of refrigeration, pasteuriza-

Death by Hamburger

On July 31, 2001, healthy two-year-old Kevin Kowal-cyk woke up with diarrhea and a slight fever. Three days later, his kidneys began to fail and his symptoms worsened. Over the following week, he was kept alive by a ventilator and a dialysis machine. Twelve days after he became ill—and to the shock of his parents and even his doctors—Kevin died. The cause? The deadly O157:H7 strain of the bacterium *Escherichia coli*,[11] almost certainly from a contaminated hamburger.

tion, and other technologies. But in some ways we are going backward. At least four aspects of large-scale industrial agriculture and food processing have increased the risk of major food-poisoning outbreaks:

- *Germs can be dispersed nationally and internationally with incredible rapidity.*[12] On a typical day, 24,000 hogs are shipped from North Carolina to 27 states, as well as to Puerto Rico, Mexico, Canada, and South America.
- *Severe crowding on industrial factory farms helps livestock-borne pathogens spread from animal to animal.* Half a century ago, a single chicken carrying a pathogen such as influenza might infect 100 others on the same farm; now that same bird might infect 50,000 others sharing its football field-sized shed. And when some mutant strains of viruses and bacteria would have only infected highly vulnerable animals, a particularly infectious agent would have died out quickly in a small flock or herd composed of mostly healthy animals. Today's huge factory farms, on the other hand, increase the chances of a germ's finding weakened animals that can act as reservoirs.
- *The widespread use of antibiotics to mitigate problems caused by crowding on factory farms adds a new dimension to food-poisoning risks.* The regular administration of low doses of antibiotics promotes the growth of antibiotic-resistant bacteria (see "Factory Farming's Antibiotic Crutch," p. 68). Thus, mutant bacteria that infect humans may be tougher to treat.
- *Industrial processing of meat allows pathogens from a small number of animals to contaminate large amounts of food.* As Eric Schlosser reminds us in *Fast Food Nation*, butchers used to provide consumers with ground beef made from a single cut. Now that large meatpacking plants have taken over, "there are hundreds or even thousands of animals that have contributed to a single hamburger," as one expert at the CDC noted.[13] Consequently,

a foodborne illness that once might have affected only one family now might affect scores of families.

Industry, of course, doesn't want to poison its customers and, under pressure from government, consumer groups, and the media, has been slowly testing and instituting new measures to prevent contamination on farms or to kill the germs at slaughterhouses. But there is a constant tension between wanting to raise and process as many animals as rapidly and cheaply as possible and ensuring that the food is as safe as possible. Compromises are always made.

Animal Pathogens Can Sicken Even Vegetarians

The hazards created by livestock production increasingly jeopardize not only the safety of meat, but also of fruits and vegetables. About 30 percent of the food-poisoning outbreaks traced to produce actually are caused by pathogens of animal origin.[14] Fruits and vegetables can be contaminated by tainted irrigation water, manure used as fertilizer, or cross-contamination from meat during transport or in the kitchen. Foods as diverse as parsley, scallions, cantaloupes, lettuce, bean and alfalfa sprouts, orange juice, and beans have caused outbreaks due to microbes characteristic of animal agriculture.[15]

While cooking kills most pathogens in meat, poultry, and some vegetables, other vegetables and fruit are not cooked. Who wants to cook one's salad to be sure it's safe?

- In lettuce plants, *E. coli* O157:H7 can be drawn up by the roots and migrate into the interior of the leaf, where the germs cannot be removed by washing. In 1996, lettuce contaminated with that bacterium caused a large outbreak of illnesses across Illinois, Connecticut, and New York. One victim was a three-year-old girl who needed surgery to remove a pool of blood from her brain and was left with damaged vision. Federal health officials discovered that cattle were penned next to the barn where the lettuce was processed and were the likely source of the contamination.[16]
- Between 1995 and 2002, 15 outbreaks were traced to *Salmonella*-contaminated sprouts. In one case, alfalfa sprouts harvested in Idaho from a field adjacent to a cattle feedlot caused outbreaks in Michigan and Virginia. The problem was so serious the U.S. Food and Drug Administration (FDA) warned in 2002 that sprouts were only safe to eat after cooking.[17]
- In 1991, *Salmonella*—presumably from animal manure on cantaloupes— caused a major outbreak, leaving a trail of illnesses across 23 states and into Canada. In 2002, *Salmonella* contaminated over 500,000 pounds of canned kale and turnip greens.[18]

Washing produce helps, but as long as animal manure is anywhere near fields and packinghouses, pathogens may be a threat.

Manure: How Many Pathogens Get Spread

Manure is one means by which germs in livestock enter the food supply and infect humans.[19] Most of the germs cause gastrointestinal problems, but *E. coli* O157:H7 causes hideously painful and sometimes fatal kidney problems (hemolytic uremic syndrome).

In a 1999–2000 USDA study of 73 cattle feedlots, 50 percent tested positive for *Salmonella*.[20] Eleven percent of samples contained *E. coli* O157:H7, and every feedlot had at least one positive sample in the course of the study.

The biggest risk to humans is probably from the fecal matter on animal hides and from intestines that contaminates meat and poultry at slaughterhouses. But pathogens in livestock manure also contaminate pools, lakes, and streams. Outbreaks of gastroenteritis (inflammation of the lining of the intestines or the stomach) traced to contaminated recreational water doubled between 1997–98 and 1999–2000.[21] *Cryptosporidium parvum* and *E. coli* O157:H7 account for nearly 90 percent of such outbreaks. A remarkable 60 percent of gastroenteritis from recreational water use occurred in *treated*

Spraying manure onto cropland also sprays bacteria.

water, such as swimming pools. If manure is not adequately treated, *E. coli* can leach into water—especially if a rainstorm occurs shortly after application to cropland—and even get into well water.[22] Of the outbreaks caused by contaminated drinking water in 1999–2000 where the cause was identified, the majority resulted from animal-borne pathogens.[23]

Farmers can compost manure to decrease the populations of bacteria enough to allow it to be spread as fertilizer, but they must control the temperature and aeration, which can be difficult and costly given the massive quantities of manure generated by large animal feeding operations. Bacteria can survive in the lagoons of liquefied livestock waste, which mimic the moist, oxygen-poor climate of the intestines in which they thrive. Thus, when lagoon liquid is sprayed onto fields, bacteria are sprayed, too.[24] Drying mounds of manure before application—another popular technique—is better than lagoon storage for eliminating bacteria, but serious risks remain. The hardy *E. coli* O157:H7 can survive for 21 months in an unaerated manure pile and for 4 months in an aerated pile. Even harsh winters cannot eradicate the germ: It can survive 100 days in frozen manure.[25]

One researcher discovered the tenacity of pathogens while studying a variety of vegetables grown in soil that was fertilized with manure inoculated with *Salmonella* and *E. coli*. Both of those bacteria were found on the harvested produce, and they also survived in the soil even after repeated cycles of freezing and thawing.[26] Furthermore, although cattle typically remain positive for *E. coli* O157:H7 for only a month, keeping a herd in a feedlot or grazing them on a field where their manure has been used as fertilizer may lead animals to be continually infected.[27]

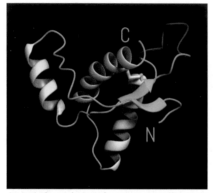

In recent years, poultry litter—ground-up feces, feathers, bedding, and spilled feed—has been fed to cattle. That practice creates a cycle that may infect those cattle with mad cow disease, because chickens are sometimes fed processed cattle products—pulverized bone and meat. If that chicken feed is excreted or spilled onto the floor by poultry, it

An artist's rendering of a prion, a small protein molecule that wreaks havoc in the brain, causing mad cow disease and the human equivalent, variant Creutzfeldt-Jakob disease.

may become part of cattle feed. The initial route through which mad cow disease was spread was the feeding of processed cattle products to cattle. So far, however, the cattle-chicken-cattle feeding cycle has not been proven to spread the disease. (For more on mad cow disease, see appendix A, p. 174.)

Diseases Direct from Livestock to You

In addition to hosting *foodborne* pathogens, farm animals carry numerous microbes that can infect people *directly*. An estimated 200 different diseases can be transferred from animals to people, and that number is growing.[28] Of 156 emerging diseases around the world, such as pfiesteria, hantavirus, and West Nile virus, 73 percent inhabit animals for part of their life cycles.[29]

Microbes from livestock can also reach people through the environment. Numerous pathogens—including antibiotic-resistant strains from livestock—are found in the air, though their impact on surrounding communities is unknown.[30] The air inside one swine barn contained *Staphylococcus, Pseudomonas, Bacillus, Listeria,* and other bacteria at worrisome levels.[31] Streams, too, could infect swimmers, boaters, and fishers. More research is needed to determine just how big a problem environmental contamination is.

The Most Threatening Animal-Borne Disease: Influenza

Influenza is the single biggest animal-borne threat, and public health officials around the globe are beginning to safeguard against possible pandemics. University of Minnesota professor Michael Osterholm warns: "Pandemics are not a question of [whether] they will happen…. The question we really have before us is how big, how bad, and when will it start."[32]

Chickens, ducks, and pigs serve as major reservoirs for flu viruses. Because pigs can become infected with both human and avian strains of a given virus, the viruses may swap genes, creating a new harmful strain to which humans may be susceptible. That process may be facilitated by mixing pigs from different farms or regions—a common event at livestock auctions or during shipping. Innocuous influenza viruses in wild birds may infect poultry, where they could undergo mutations that enable them to infect and kill humans. The gravest risk arises when flu viruses gain the ability to spread directly from person to person.[33]

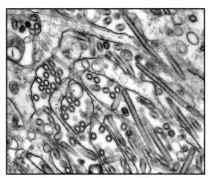

Avian influenza virus A H5N1 (gold) is growing in cultured cells (green).

Various gradually changing strains of influenza virus are endemic and cause annual nationwide outbreaks in the United States. In an average year, 10 to 20 percent of the population gets the flu, with 114,000 requiring hospitalization and 36,000 dying.[34] Of course, those figures are dwarfed by the massive 1918–19 flu pandemic, which killed more people faster than any disease ever.[35] While "only" 500,000 Americans died, some countries lost half their populations.[36] Globally, as many as 50 million people died. That strain of flu likely came from birds and then spread to humans. If a similar strain of flu struck today, some experts estimate that 1.8 million Americans would die.[37]

Poultry-related influenza outbreaks have been much in the headlines in recent years. In 1997 in Hong Kong, a strain of avian influenza ("bird flu") H5N1 leapt from poultry to humans, infecting 18 people.[38] Six people

Factory Farming's Antibiotic Crutch

Food poisoning is bad enough when you're infected with ordinary germs. But when those germs are resistant to customary antibiotics, ordinary illnesses may become life threatening. We're courting disaster when we allow farmers to use penicillin, erythromycin, and other important antibiotics for economic—not medical—reasons.

Antibiotics, the first true miracle drugs, have saved countless lives over the past half-century. But far greater quantities of antibiotics are used in farm animals than in humans.[39] The drugs are sometimes used to treat sick animals, but mostly they are administered at low, non-therapeutic levels to whole flocks and herds to promote growth and counteract the dirty, crowded conditions in which most animals are raised.

Antibiotic Use Breeds Resistance

Using low levels of antibiotics day in and day out on millions of animals greatly increases the chances that bacteria—including those that cause foodborne illnesses—will develop antibiotic resistance. The problem arises when a germ happens to mutate in one of several ways that reduces the antibiotic's effectiveness. The tougher new bacteria:

- pump the antibiotic out of their cells,
- degrade the antibiotic,
- change the antibiotic's chemical structure, or
- modify target molecules to "fool" the antibiotic.

The antibiotic kills off all but the resistant germs, which then flourish. If people are infected by those bacteria via contaminated food, they can suffer illnesses that may only be cured by the newest, most powerful (and expensive) antibiotics. Farmers and others in direct contact with livestock can also be infected by the resistant bacteria.[40]

The U.S. Department of Health and Human Services has recognized that "Antimicrobial resistance among foodborne bacteria, primarily *Salmonella* and *Campylobacter*, may cause prolonged duration of illness, and increased rates of bacteremia (bacteria in the blood), hospitalization, and death."[41] Antibiotic-resistant *Salmonella*, a common foodborne pathogen, causes at least 29,000 extra illnesses, 342 extra hospitalizations, and 12 extra deaths per year.[42] The ultimate danger is that bacteria will develop resistance to all the common antibiotics and cause a deadly epidemic.

A 2001 U.S. Food and Drug Administration study of ground meat and poultry found that 20 percent of the samples contained *Salmonella*, and over half of those bacteria were resistant to at least three important antibiotics.[43] Even more alarming, some strains of *Salmonella* and other foodborne pathogens were resistant to a dozen different antibiotics. The livestock industry's profligate use of antibiotics almost certainly selects for those "superbugs."[44]

In 1995, the FDA—over the objections of the Centers for Disease Control and Prevention—allowed chicken farmers to treat whole flocks with fluoroquinolones, a family of powerful new antibiotics, even if only a few birds were sick. Predictably, rates of resistance in *Campylobacter* quickly soared from virtually zero to 20 percent.[45] That spurred the FDA, in 2000, to reverse course and propose barring flock-wide use of fluoroquinolones.[46] Two years later, the agency estimated that fluoroquinolone-resistant infections were causing over 17,000 additional cases of food poisoning, leading to 95 hospitalizations.[47] Only two companies marketed the antibiotics: Abbott Laboratories immediately stopped marketing its product, but it took five years to overcome Bayer Corporation's opposition and to stop farmers' use of its similar drug.[48]

Growing Opposition to a Dangerous Practice

Livestock producers and the animal-drug industry insist that giving animals low doses of antibiotics is safe.[49] But public health experts counter that it is senseless to endanger the effectiveness of vital human medicines—especially when they are not essential to farmers. The American Medical Association, American Public Health Association, and other health groups have opposed unnecessary uses of antibiotics on farms. The American Academy of Pediatrics found that "children are at an increased risk" from antibiotic-resistant infections rooted in non-therapeutic uses of antibiotics in food-producing animals. And a study by the Institute of Medicine concluded that the "FDA should ban the use of antimicrobials for growth promotion in animals if those classes of antimicrobials are also used in humans." The World Health Organization made a similar plea. More than 300 local and national organizations, including the medical, public health, and pediatrician organizations mentioned above, have supported legislation to limit the use of antibiotics in livestock.[50]

Antibiotics are widely used in crowded, dirty animal facilities to prevent or treat bacterial infections.

Industry maintains that antibiotics help healthy animals grow faster and at a lower cost. But a committee of the National Academy of Sciences emphasized that the "beneficial effects of subtherapeutic drug use are found to be greatest in poor sanitary conditions."[51] Just as public health experts finally figured out that cleaning up the water and the air drastically reduced infectious diseases in people, so agribusiness should look to use different approaches to prevent illnesses in their animals. If they cleaned up their hog sheds, gave their chickens more room to roam around, stopped feeding cattle an unnatural grain-rich diet, and bred animals not just to grow fast but to have strong immune systems, farmers could both raise healthier animals and protect the effectiveness of precious antibiotics.

The European Union began phasing out the use of medically important antibiotics in healthy animals in 1999 and banned that use completely on January 1, 2006.[52] Denmark, the world's largest exporter of pork, moved even faster. In 1998 it instituted a virtual ban (through a $2 tax on treated pigs) on using growth-promoting antibiotics in pigs after weaning.[53] In 2004, farmers were not using any antibiotics to promote growth, though more antibiotics were being used to treat illnesses. The total poundage used is dramatically lower than before the ban, and the prevalence of both resistant and nonresistant foodborne pathogens plummeted in hogs and their meat.[54] Moreover, Danish economists estimate that the cost of producing pork will rise just 1 percent.[55]

Change is coming, if more slowly, in the United States.[56] Tyson Foods, the nation's largest chicken producer, reduced its use of antibiotics by 93 percent between 1997 and 2004, and three other major companies say they have stopped using antibiotics on healthy animals. The Iowa Pork Producers Association is now urging "all Iowa pork producers to voluntarily discontinue use of all growth-promoting antibiotics" in the feed of pigs that weigh more than about 50 pounds. And a rapidly growing number of organic livestock producers do not administer any drugs at all (they treat sick animals, but then do not market them as organic). Probably reflecting such developments, between 1999 and 2004 the volume of antibiotics used in animals declined by 10 percent, despite a 5 percent increase in livestock production.[57] Unfortunately, there is no similar progress in the cattle industry.

died—a fatality rate of 33 percent. Hong Kong officials responded by ordering the slaughter of 1.4 million birds. Luckily, the disease did not spread easily from person to person, so control measures were effective. Since 1997, however, four more outbreaks of avian influenza have occurred in Hong Kong, prompting the government to respond with such preventive measures as poultry vaccinations and new restrictions on imported poultry. Between 2003 and 2006, bird flu spread to other parts of Asia and countries in Europe and Africa. It has killed over 100 people and prompted the slaughter of more than 150 million poultry, costing the industry billions of dollars.[58]

The CDC says that "The avian influenza…outbreak in Asia is not expected to diminish significantly in the short term."[59] In 2004 in North

America, a milder strain of avian influenza emerged in Canada and Texas. No human deaths were reported, although two poultry workers became ill.[60] Some 17 million chickens, turkeys, and ducks were culled to prevent the virus from spreading. In 2006, veterinary and health experts in North America and elsewhere were bracing for a new round of infections. Tara O'Toole, director of the University of Pittsburgh Medical Center's Center for Biosecurity, speculated that a highly infectious bird flu virus could kill as many as 40 million Americans.[61] While the most dire predictions are likely overblown—partly because mutations are expected to weaken the virus if it "learns" to spread from person to person—the possibility of epidemics is enormously enhanced by the widespread raising of large numbers of livestock.

In huge poultry sheds, germs from one bird can easily infect thousands of other birds.

Weak Safeguards Endanger Consumers

All of the problems mentioned above are exacerbated by the federal government's incomplete and fragmented food-safety system. For starters, the United States does not have a system that tracks animals and meat from the farm to the slaughterhouse to the table. That prevents health officials from tracing the cause of a food-poisoning outbreak back to the farm. Also, the government cannot require food processors to recall products that are suspected of causing outbreaks; instead, they must ask and negotiate with companies—while people are getting sick. The USDA cannot fine companies for violating the law, and the FDA can only fine a company $1,000 and threaten officials with a year in jail. Those agencies' real power comes from their authority to seize products on store shelves and generate bad publicity. As for imported foods, the USDA has the power to inspect foreign processing plants, but the FDA does not.

Most of the responsibility for ensuring a safe food supply rests with the USDA and the FDA, with almost a dozen other agencies playing smaller roles. The USDA oversees the safety of meat, poultry, pasteurized eggs, and processed foods containing meat or poultry, while the FDA oversees everything else, including produce, eggs in their shells, seafood, and processed foods that contain little or no meat or poultry. That division creates some bizarre situations. For example, the USDA regulates dehydrated chicken soup, but the FDA oversees dehydrated beef soup. Peculiarly, though, the FDA regulates chicken broth, but the USDA regulates beef broth. (The government is looking to correct that particular bit of bureaucratic craziness.)

USDA microbiologists obtain samples for microbial analysis from a washed carcass.

More importantly, federal funding priorities are misguided. CSPI's food safety director Caroline Smith DeWaal emphasizes that while FDA-regulated foods cause two-thirds of all outbreaks, the FDA receives only 38 percent of food-safety funding. As a result, that agency performs too few inspections of the facilities it oversees. The USDA inspects meat and poultry plants daily; the FDA inspects other operations only about once every five years on average.[62]

What It All Means

Animal products cause many foodborne infections in the United States, and livestock are the source of other infectious diseases, such as the flu, that are spread by vehicles other than food. Sicknesses and deaths aside, those illnesses generate enormous health-care and other costs. Some of the production systems that animal agriculture uses promote the spread of dangerous pathogens from animals to meat to humans and from animal manure to fruits and vegetables. Industry is well aware of the food-safety problem and has been attacking it with new technologies, ranging from steam-treating and acid-washing beef carcasses to vaccinating poultry to irradiating cuts of meat. Still, foodborne and farm animal–related illnesses likely will never be eliminated totally. Meanwhile, the government's food-safety system, which includes programs that are perpetually underfunded and riddled with holes, has proved inadequate in fulfilling its public health mission. With a large percentage of foodborne illnesses caused by animal products, one personal solution is obvious: eat fewer animal products—and wash your fruits and vegetables.

Better Soil

"Soybean production is killing us," notes Larry Gates of the Minnesota Department of Natural Resources. Southeast Minnesota, which once boasted clean rivers and streams, is increasingly inhospitable to healthy and diverse aquatic life—as well as to the people who flocked to those waters to fish and swim. Encouraged by Farm Bill incentives, Minnesota farmers have been converting their pastures and grasslands to soybean fields. That simple switch has had a profound impact, as endless rows of soybean plants have led to unprecedented levels of erosion. Load upon load of sediment has been washed into the river. As a result, brown trout populations, which had been rising for decades, are declining to the point where hundreds of thousands of young trout will have to be placed in the river if the population is to be maintained.[1]

Producing food animals, and the grains and soybeans that speed their growth, takes a tremendous toll on farmland—particularly its precious topsoil. Growing crops for animal feed frequently erodes the soil, as does overgrazing of grasses by livestock. Further, cattle's constant trampling of vulnerable rangeland can almost irreparably damage the environment. The immense quantities of fertilizers—including old-fashioned manure, urban processed sewage sludge, and conventional chemicals—and pesticides used to grow feed grains contain nutrients and toxins that disrupt the soil ecosystem, poison wildlife, and pollute local and far-off waterways.

- Raising almost 100 million acres of feed crops for livestock production depletes topsoil of nutrients and causes erosion.

- About 22 billion pounds of fertilizer—about half of all fertilizer applied in the United States—are applied to lands used to grow feed grains for American livestock annually. The energy needed to manufacture that fertilizer could provide a year's worth of power for about 1 million Americans.

- Livestock may damage the land they graze on by compacting the soil, making it difficult for the soil to absorb water.

- Soil—and crops—can be contaminated with cadmium, lead, and other heavy metals in sewage sludge and chemical fertilizers.

Agriculture has an enormous impact on soil and soil quality: Grazing land and cropland are the second- and third-largest uses of land in the United States (forests are the largest), together accounting for just under half of America's total acreage.[2] In contrast, urbanization and sprawl affect only about 3 to 5 percent of the U.S. land area.[3]

Importance of Good Topsoil

Soil, along with water and sunlight, is one of the three fundamental elements of crop production. A thick layer of topsoil, rich in such nutrients as nitrogen, phosphorus, and potassium, absorbs and holds rainwater well and provides the best environment for growing crops.

But topsoil can be lost, leached away by water or blown away by wind. The U.S. Department of Agriculture (USDA) estimates that almost 2 billion tons of topsoil eroded from cropland in 2001.[4] That's a huge amount, but represents a 40 percent decline since 1982. The main cause of erosion is the lack of plants that hold the soil in place. Native meadow grasses, hay, and small grains such as wheat help protect topsoil by providing a solid cover over a field.[5] Many large farms, however, plant livestock feed crops, such as corn and soybeans, that are grown in rows and endanger topsoil since the bare patches between each row are relatively susceptible to erosion. The loss of topsoil reduces fertility,[6] which increases the need for chemical fertilizers. And the switch from healthy natural topsoil to artificial nutrients leads to a whole host of problems—nutrient imbalances, runoff, and water pollution—detailed later in this chapter.

Livestock's Demand on Soil

Feeding grain to livestock and then eating the livestock (or their eggs or milk) needs a lot more land than just eating the grains themselves. Raising livestock creates a huge demand for corn, soybeans, and a few other crops. About 66 percent of U.S. grain ends up as livestock feed at home or abroad.[7] While pigs and chickens consume a good share of that grain, cattle at feedlots are the biggest consumers, in part because they are the least efficient converters of grain to meat. Outside the United States, livestock consume only 21 percent of total grain production, with the vast majority of grain consumed directly by people. But as nations' incomes rise, so does their appetite for pork, chicken, and grain-fed beef.

It's much more efficient in terms of land, water, and other resources for people to eat grains, such as the wheat grown on this Utah farm, than for people to eat foods from animals that ate the grain.

Frequently, farmers respond to the huge demand for feed grains by turning to monocropping—raising single crops over huge areas—or they use limited rotations, where two crops destined for livestock feed are raised in alternating years. About 16 percent of corn—over 12 million acres—is raised without any rotation at all, though the majority of corn—59 percent—is rotated with soybeans.[8] Meadow grasses and small grains (such as wheat), both vital to the preservation of topsoil, are included in only 8 percent of corn rotations, according to the USDA.[9]

Good soil health depends on several factors, including maintaining nutrient and organic matter content and avoiding topsoil loss.[10] Robust crop variation—including seasons when land remains fallow altogether—is critical to maintaining optimal soil health. Including soybeans in a rotation helps maintain nutrient levels because soybeans and other legumes can "fix" nitrogen (the process by which bacteria convert nitrogen from its relatively inert gaseous form in the atmosphere into compounds useful as nutrients, such as nitrate). However, soybeans, because they leave little residue on the field after harvest, are even less protective of topsoil loss than corn.[11]

Erosion

A typical acre of U.S. cropland loses 5 tons of soil each year.[12] About 20 percent of cropland—some 65 million acres—erodes at a rate that actually decreases its productivity.[13] The resulting nutrient losses and lowered yields cost almost $10 billion per year (see table 1). And soil's reduced water-holding capacity is not only costly (an estimated $3.2 billion per year) but self-perpetuating. It increases the rate of further erosion because unabsorbed water flows over the soil, with less water remaining for plants. Eroded soils therefore likely need more irrigation than "healthy" land—but irrigation, in turn, promotes more erosion.

The problems caused by cropland erosion extend well beyond the farm. Soil carried away by wind creates dust and haze and causes respiratory illnesses and property damage, which together cost over $14 billion per year.[14] Impaired water quality, due to sediment damage from agricultural runoff, accounts for about one-third of the cost of erosion. When soil is deposited into water, the suspended particles block sunlight, impairing the growth of

Eroded Soil, Eroded Yields

Row crops such as corn and soybeans are vulnerable to erosion because of the naked patches of land that lie between the rows. Some soil-building innovations, such as planting cover crops after the main crop has been harvested,[15] almost keep pace with erosion, but they are not universally used.

Comparing cropland to other land uses demonstrates how damaging row-crop production is to topsoil. Erosion reduces the productivity of more than 20 percent

of cropland. That compares to only 6 percent of private pastureland—or fewer than 8 million acres.[16] Because responsibly grazed pastureland typically has limited exposed soil, only about 1 ton of soil is lost per acre of pasture per year, in contrast to the 5 tons for cropland.

Table 1. **The cost of erosion on all U.S. cropland (2004 $)**[17]

Location	Problem	Cost per ton of eroded topsoil	Total cost per year (billions)
Cropland	Nutrient losses and reduced yields	5.16	9.8
	Reduced water-holding capacity	1.69	3.2
Offsite (off cropland)	Impaired water quality	7.44	14.1
	Property damage	3.89	7.4
	Health effects from air pollution	3.72	7.1
Total		$21.90	$41.6

aquatic plants and depriving animals that feed on them of food. Sediment can also raise water temperatures, disrupting the habitats of aquatic species. But perhaps the greatest harm is not from the soil itself, but from fertilizers and pesticides that attach to soil particles.[18] The cost of water pollution from erosion is estimated at $14 billion per year—and that doesn't take into account the health and environmental harm from runoff from agricultural chemicals.[19]

"Erosion is one of those problems that nickels and dimes you to death: One rainstorm can wash away 1 millimeter of dirt. It doesn't sound like much, but when you consider a hectare (2.5 acres), it would take 13 tons of topsoil—or 20 years if left to natural processes—to replace that loss.... Yet controlling soil erosion is really quite simple: The soil can be protected with cover crops when the land is not being used to grow crops."
—David Pimentel, professor of ecology, Cornell University[20]

Compaction

Compaction occurs when topsoil—particularly when it is wet—is subjected to the intense weight of the heavy machinery farmers use to cultivate, plant, and harvest fields and of large livestock such as cattle—though machinery typically is the more damaging.[21] Compaction makes soil too dense for plant roots to penetrate easily, reducing the rates of plant growth and crop yields.[22] It also reduces soil's ability to absorb water. The American Society of Agricultural Engineers found that pasture grazed by cattle for 10 years absorbs less than one-fifth as much water as ungrazed pasture.[23] One consequence of compaction is erosion, because water that is not absorbed runs off, carrying topsoil with it.

Soil compaction is a major problem on western rangelands where cattle congregate in the biologically rich areas along the banks of waterways or in wetlands. That compaction reduces the capacity of those wetlands and soil to hold water, which leads to greater flooding and inhibits the recharging of water tables.

Compaction poses a different, but not a lesser, problem in the arid and semiarid regions of the West. Few grasses, bushes, and other plants grow on these lands. Instead, the main soil covering is an interconnected community—collectively referred to as *microbiotic crust*—of mosses, lichens, and cyanobacteria. (This last is an unusual form of bacterium that uses chlorophyll and other pigments to capture light for photosynthesis.) Crusts help hold soil nutrients, control water absorption, and create a medium for plant growth. Although tough enough to support life in some of the hottest, driest climates in the United States, crusts are quite vulnerable to physical disturbances. Because the crusts are only 1 to 4 millimeters thick (less than one-sixth of an inch), compaction and grazing by cattle can easily destroy them. And that destruction inevitably leads to erosion, water loss, and harm to native plant species. Moreover, crust recovers extremely slowly. Full regeneration takes 50 to 250 years, depending on the extent of damage, according to government scientists.[24]

Another Problem: Exotics

Heavy grazing by livestock promotes the spread of exotic, invasive weeds. Those plants provide less-suitable land cover and do not hold soil together as well as native plants. Cattle contribute to the spread of such weeds in three ways:

- They graze on native species, ignoring exotic weeds, which can then proliferate.
- They spread the seeds of exotic plants.
- Trampling by animal hooves makes ideal seedbeds for exotic plants.[25]

New Practices Help, but More Help Is Needed

Over the past two decades, farmers have used various measures to better conserve farmland. And that has paid off: In 1982, 3 billion tons of topsoil eroded from cropland. By 1997, that figure was reduced by 40 percent to just under 2 billion tons.[26] But in some areas, soil losses remain well above levels of sustainability.

Several factors account for the dramatic improvement in soil conservation. For starters, the USDA's Conservation Reserve Program (CRP) has paid tens of thousands of farmers to idle their most erodible lands, thereby dramatically improving soil health.* The CRP idles about 35 million acres of land.[27] Only about 1 percent of CRP land, fewer than 1 million acres, is eroding at an unsustainable rate.[28,†] That success is impressive, particularly since most of the land included in the program was experiencing serious erosion. The CRP shows that even in extreme cases, strong (though expensive) measures can protect the land.

Farmers also have reduced erosion by using *conservation tillage* or *reduced tillage* on roughly half the nation's cropland. That practice cuts back on plowing

and leaves crop residue (such as cornstalks) on the ground after harvest to prevent erosion.[29] "No-till" agriculture, which is facilitated by genetically engineered herbicide-tolerant soybean and corn varieties, barely disturbs soil from planting to harvest time.[30] Farmers also have been planting buffer strips or terracing land to help reduce erosion.

No-till soybean crops minimize soil erosion.

Topsoil losses persist nonetheless. Reducing or eliminating the need for corn, soybeans, wheat, and other grains for livestock feed—especially for cattle—could further reduce erosion. In theory, ceasing

---

*CRP land may be grazed or cut for hay under emergency conditions such as drought or an animal feed shortage, but it otherwise remains fallow.

†Though the vast majority of acres enrolled in the CRP are "highly erodible land," other lands are also enrolled to protect wildlife habitats and water quality and to address other environmental problems. Inclusion of those acres lowers the average rate of erosion on CRP land.

Wind erosion still occurs on about 420,000 acres of CRP land and water erosion on 365,000 acres, with some land experiencing both types of erosion. However, even if there were no overlap, only 2.4 percent of all CRP land would experience erosion-induced productivity losses.

grain production for livestock would allow close to 100 million acres to lie fallow and revert to natural grasslands and woodlands.[31] That shift could save as much as 700 million tons of topsoil per year. In reality, though, much of that land would be used to grow crops for export or for conversion to gasohol, high-fructose corn syrup, and other products, and

Alternating strips of alfalfa with corn on the contour helps reduce soil erosion on this Iowa farm.

some would be planted in crops that would replace some of the meat in our diet.

Effects of What We're Putting on the Soil

Loss of topsoil decreases productivity, so to compensate for that farmers add soil nutrients. That means applying fertilizer—and lots of it—in the form of chemicals, manure, or treated sewage sludge.

Chemical Fertilizers

Fertilizer causes environmental problems primarily because farmers often apply too much to their land. Because about half of all fertilizer applied

in the United States is used solely for raising feed grains for animals, reducing that usage could reduce environmental degradation.[32]

Even when not over-applied, nitrogen fertilizer causes serious environmental problems. That fertilizer is usually applied as ammonium nitrate, which can react with oxygen in the air and release ammonia. Ammonia can damage local ecosystems, including the plant life on the fertilized land.[33] When carried by wind and rain, the ammonia may be deposited in waterways and affect distant ecosystems (see "Ammonia," p. 104, for further details).

Fertilizer Used to Produce Meat, Poultry, Eggs, and Milk

Producing different animal products requires very different amounts of fertilizer.[34] In all, 22 billion pounds of fertilizer are used per year.

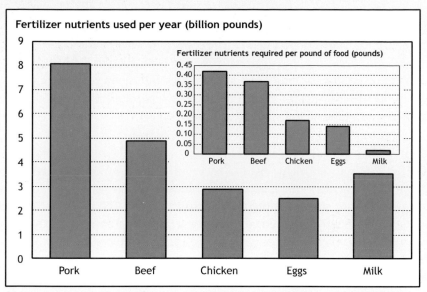

Fertilizer nutrients used per year (billion pounds)

Fertilizer nutrients required per pound of food (pounds)

Notes: Inset chart is for cooked food, except for milk. Data exclude exported crops and food.

- Hogs are the least fertilizer-efficient of major farm animals, partly because, unlike cattle, they eat grains their entire lives. It takes about a pound of fertilizer to produce 2½ pounds of cooked pork.

- Producing beef requires large amounts of fertilizer, in large part because cattle are inefficient converters of feed to meat. One pound of fertilizer is needed to produce 3 pounds of cooked beef.

- Chicken and egg production require less than half as much fertilizer per pound as beef or pork.

When the oxygen content of soil is low, nitrogen fertilizer undergoes a process called *denitrification*, which yields a variety of nitrogen-containing gases, including nitrogen gas, nitric oxide and nitrogen dioxide (which are together known as NO_x, since, in the presence of sunlight, they rapidly interconvert), and nitrous oxide.* The harmless nitrogen gas simply returns

---

*The fact that oxygen-containing species are produced may seem strange considering that the reaction takes place in the absence of oxygen. What actually happens is that in oxygen-poor conditions, anaerobic bacteria strip the oxygen from nitrogen dioxide (which occurs naturally in soil or is deposited by acid rain), releasing nitrogen gas and nitric oxide into

to the atmosphere. However, NO_X destroys ozone, impairs lung function, and contributes to fog and acid rain.[35] It also travels even farther from its source than ammonia.[36] Nitrous oxide is a destructive greenhouse gas 300 times more potent than carbon dioxide (for more on this topic, see "Nitrous Oxide" and "Nitric Oxide and Nitrogen Dioxide," pp. 107 and 108).[37] Agriculture contributes about 37 percent of all nitrous oxide releases in the United States, with much of that coming from fertilizer.

Besides polluting the air, fertilizers also increase the acidity of soil.[38] That reduces the soil's ability to hold nutrients and can permanently reduce soil productivity. Acidification ordinarily is controlled by applying even more chemicals, such as lime (calcium carbonate).

Heavy Metals in Chemical Fertilizer

The potash and phosphate ores used to produce chemical fertilizers frequently contain heavy metals that may contaminate the soils on which they are used. Those contaminants can be absorbed into the grains grown in the soil, the livestock that consume those grains, and eventually the people who consume the resulting meat and dairy products.[39] The U.S. Environmental Protection Agency recognizes that cadmium, lead, arsenic, zinc, and other minerals sometimes contaminate fertilizer.[40] With intensive application of nitrogen, phosphate, and potassium fertilizers, cadmium and lead levels in soil can double in a dozen years.[41] Liming materials, such as sludge from water treatment facilities (see "'Biosolids' Fertilizer: Processed Sludge," p. 84), also contain a potpourri of heavy metals, including mercury.[42] So when liming materials are used to reduce the acidity of soil, they also may pollute it.

A 1999 study of toxic waste in California by the nonprofit Environmental Working Group found that one in six samples of commercial fertilizers exceeded the state's criteria for what constitutes hazardous waste. Among the heavy metals detected, lead and arsenic were present in the greatest amounts.[43]

The concentrations of metals may be even greater in manure than in chemical fertilizers, and transferring them to soil may lead to higher levels in food crops.[44] In fact, many poultry farmers add to feed an arsenic-containing drug, roxarsone, to kill parasites that slow the animals' growth. The U.S. Geological Survey states that each year the poultry litter that is spread onto nearby fields contains 2 million pounds of roxarsone and "could result in localized arsenic pollution."[45] Johns Hopkins University

---

the atmosphere. Oxygen in the atmosphere readily recombines with those gases to produce nitrous oxide and NO_X.

Manure—Excess of Riches

In 2000, livestock in the United States produced about 3 trillion pounds[46] of manure (including feces, urine, and poultry litter). That's 10 times as much as people produced.[47] Cattle accounted for about three-fourths of the manure (see figure).[48]

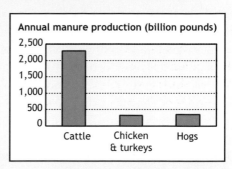

Annual manure production (billion pounds)

In 1997, farms produced 1.5 billion pounds more manure nitrogen and almost 1 billion pounds more manure phosphorus than could be used on fields.[49] However, much cattle manure is deposited harmlessly (or beneficially) on pastureland.

Farmers typically deal with that over-abundance of manure by spraying it on nearby fields as fertilizer. That adds organic matter to soil, increasing the soil's water-holding capacity and fertility. It also spares the considerable resources needed to manufacture chemical fertilizers.[50]

But using manure as fertilizer has severe limitations. For starters, the nutrients occur primarily in an organic form and are relatively unusable by plants, which prefer inorganic nutrients. Also, manure may be difficult to collect, expensive to store and transport, and not have the desired proportions of nutrients.[51] The University of Maryland Agricultural Extension Service states, "Typically if a farmer uses manure to fulfill a crop's nitrogen requirements, he is overfertilizing for phosphorus and potassium. If manure is used to meet phosphorus or potassium requirements, additional nitrogen will be required from other sources."[52] Also, the release of nutrients from manure to the soil cannot be timed to match the needs of plants, as it can with chemicals.

Faced with mountains of manure from intensive feeding operations, researchers are exploring new solutions. Dried manure can be burned as an energy source, be added to aquaculture ponds to induce algal growth (which would provide food for fish), or even be used as a building material. Of course, one obvious way to reduce the 3-trillion-pound annual load of animal manure would be to reduce animal populations and rely less on cattle feedlots. Eating less meat, especially from animals raised in confinement, would encourage farmers to do that.

researchers warn that "If animal waste were classified as hazardous waste, it would be prohibited from land disposal based solely on its concentrations of leachable arsenic."[53]

"Biosolids" Fertilizer: Processed Sludge

To address their waste-disposal problems, cities sell treated sewage sludge—biosolids—to farmers cheaply as fertilizer. Sixty percent of processed urban sewage sludge—3.4 million tons per year—is now applied as fertilizer. In theory, that approach is mutually beneficial, because it enables cities to dispose of their waste, while providing farmers with affordable fertilizer. The one problem—and it's a significant one—is that sewage can be tainted with industrial waste and pathogens.[54] Government regulations are supposed to restrict levels of heavy metals; volatile organic chemicals; and pathogenic bacteria, viruses, and parasites.[55] But the controls sometimes fail. In 2003, hundreds of cows at Georgia dairy farms died after they ate hay grown on fields fertilized by processed sewage sludge.[56]

Currently, fertilizer manufacturers are not required to disclose heavy-metal content on product labels, so the full extent of the problem is unknown. To date, only Washington and Texas limit heavy-metal contaminants (including those from industrial sludge) in fertilizer.[57]

Pesticides: Gauging the Health Risk

Large amounts of pesticides—and potentially dangerous (and misnamed) inert chemicals included in pesticide products—continue to be applied to soil, though the current volume is 40 percent less than was used in the late 1970s and early 1980s.[58] Pesticides can unintentionally harm plants and animals; organisms living in the soil; and fish and other animals, plants, and microorganisms in the waterways into which the chemicals are carried. Because they adhere to particles in soil, pesticides can be carried long distances on dust and then tracked into homes and public spaces.

Glyphosate (marketed under the name Roundup) and atrazine are the two most widely used herbicides, helping control weeds on millions of acres of soybeans, corn, and other crops. Over 100 million pounds of those two pesticides are used every year. Even though their half-lives are moderate (between 30 and 100 days, depending on environmental conditions[59]), significant residues still may be present in soil after a year.

Because of their widespread use, scientists have explored the possible environmental and health effects of glyphosate and atrazine. Both have been implicated in the decreases in amphibian populations seen in the upper Midwest and elsewhere around the world. University of Pittsburgh researchers have discovered that a supposedly inert ingredient in glyphosate endangers amphibians.[60] Rick Relyea and two colleagues studied the detergent (polyethoxylated tallowamine) that helps glyphosate get into plant leaves. At doses that are likely to occur in nature, the detergent kills tadpoles and frogs. Relyea considers Roundup "extremely lethal to amphibians."

The gray tree frog (on top) and American toad (on bottom) are both harmed by an ingredient in the herbicide Roundup.

Atrazine, used by most corn farmers, also affects amphibians. Tyrone Hayes and his colleagues at the University of California at Berkeley exposed frogs to levels of atrazine lower than what is permitted in drinking water and found that the herbicide caused gonadal and limb abnormalities and hermaphroditism.[61] Hayes uses the term "chemical castration," and says, "because the hormones that are being interfered with occur in all vertebrates, maybe they're telling us it's just a matter of time" before atrazine is found to harm humans.[62]

Pesticides eventually are broken down in the soil by microorganisms or through chemical reactions, or they are carried into groundwater or streams. Some of the harm they can cause there is discussed in "Pesticides Wash Off of Farmland," p. 100.

What It All Means

Healthy topsoil is crucial to producing crops, but modern agriculture has placed extraordinary demands on cropland. The enormous quantities of feed grains that farmers produce help satisfy our desire for inexpensive meat and dairy products—but at great cost to topsoil, the environment, and even human health. The row crops that stretch from one end of the horizon to the other in many parts of the United States provide less anchorage for topsoil, increasing erosion. The chemical and biosolids fertilizers applied to farmland sometimes upset the balance of nutrients, as well as release into the atmosphere gases that harm human health and the environment. And the pesticides applied to the land and crops disrupt ecosystems, harm wildlife, and—as discussed in "Risks from Pesticides," p. 53—endanger farmworkers and possibly consumers.

Argument #4.
More and Cleaner Water

Lake McConaughy, which receives most of its water from the North Platte River, was once considered "Nebraska's ocean" and was a haven for migrating eagles and other birds. But times have changed. After years of heavy irrigation by farmers raising animal feed grains such as soybeans and corn, fully half the water the lake can hold has been lost, especially during dry summers. Consequently, water supplies for hydroelectric power are on the wane, and the Central Nebraska Public Power and Irrigation District, the lake's owner, is severely rationing water for farmers and ranchers. Although that might help future conditions, irrigation has, in the words of local resident Ruth Clark, taken "a beautiful, majestic lake and turned it into a mud hole."[1]

Raising livestock requires enormous amounts of water. Although the United States is blessed with water supplies far exceeding consumption, water is not distributed evenly throughout the country. In large swaths of the West, demand from farmers who want to irrigate their crops and the thirst of soaring urban populations often outstrip the supply. Cities

such as Albuquerque, Denver, and Phoenix, all of which draw water from the Colorado River, face water shortages and water-quality problems due to local farmers (who were there first).[2] Farms, especially those growing feed grains and cotton and raising livestock, are using up groundwater and surface water—permanently. At the same time, those farms cause soil erosion and dump fertilizer, manure, pesticides, and topsoil into nearby rivers and streams. The end result in some places is water so polluted it is unsafe to drink and uninhabitable by various aquatic animals.

- Agriculture uses about 80 percent of all freshwater in the United States.
- It takes about 1,000 gallons of irrigation water to produce a quarter-pound of animal protein.
- Half of all irrigation water is used to raise livestock. About 14 trillion gallons annually water crops grown to feed U.S. livestock; another 1 trillion are used directly by livestock.
- The water used to irrigate just alfalfa and hay—7 trillion gallons per year—exceeds the irrigation needs of all the vegetables, berries, and fruit orchards combined.
- Farms pollute water with fertilizer, pesticides, manure, antibiotics, and eroded soil.

The Water Cost of Meat Production

Producing meat takes large amounts of water (see figure 1). The animals themselves need water to drink and to cool themselves, and farmers need vastly greater amounts of irrigation water to grow the grains and roughage that are fed to the animals. An average of about 1,000 gal-

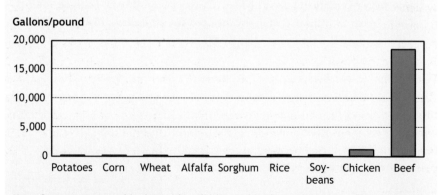

Figure 1. **Water used to produce various crops, chicken, and beef**[3]

Gallons/pound

Note: Crops are expressed in dry weights. Chicken and beef are adjusted to edible portion; our adjustment assumes that 28 percent of beef cattle and 39 percent of chicken is edible. Figures include water from rain and irrigation.

lons of irrigation water are needed to produce 1 pound of animal protein (more for beef, less for poultry). That irrigation water is supplemented by larger amounts of rainwater, especially in big corn-growing states such as Illinois and Iowa.[4]

Together, irrigating feed crops and raising livestock consume over half of all freshwater (see figure 2).[5] In contrast, domestic uses—all showers taken, toilets flushed, cars washed, glasses drunk, and lawns watered—consume less than one-tenth as much water as agriculture.

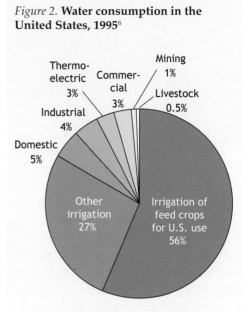

Figure 2. **Water consumption in the United States, 1995**[6]

Total: 37 trillion gallons/year

Irreplaceable Groundwater Is Being Depleted

About 90 percent of U.S. water is renewable, coming from rain, lakes, and rivers. The remainder largely is from nonrenewable underground aquifers (groundwater).[7] Agriculture accounts for about 80 percent of all freshwater consumption in the United Sates and over 60 percent of groundwater use.[8]*

Nationally, though many aquifers get recharged, the overall rate at which water is removed from aquifers exceeds the rate of replenishment by as much as 21 billion gallons *per day*.[9] In the largest and perhaps most severely depleted aquifer—the Ogallala, which underlies parts of Texas, New Mexico, Oklahoma, Kansas, Colorado, Nebraska, South Dakota, and Wyoming—water levels are falling several inches per year.[10] The Ogallala Aquifer is 1,000 feet deep in some parts of Nebraska, but in some parts of the Great Plains, it has dropped from 230 feet deep to only about 20 feet over the past 25 years.[11] The majority of water extracted from the Ogallala is used to irrigate crops.[12] Some farmers who depend on it may be facing high prices or dry wells in coming years.

---

*Both the U.S. Department of Agriculture and U.S. Geological Survey estimate water usage, but they use different measuring techniques and report somewhat different amounts. This chapter uses figures from both agencies as noted in the text and endnotes.

When an aquifer shrinks in coastal areas—including those with farms nearby—saltwater replaces groundwater. That permanently diminishes the aquifer's value.[13] Additionally, the loss of underlying groundwater sometimes causes land subsidence, a sinking of the Earth's surface. Land subsidence has affected more than 17,000 square miles in 45 states—an area twice the size of New Jersey.[14] According to a 1991 estimate from the National Research Council, land subsidence causes flooding and damage to buildings, roads, and other structures, with the cost amounting to over $125 million per year.[15]

Irrigation Water: Trillions of Gallons Wasted

American farmers irrigate about 56 million acres of land, or 88,000 square miles.[16] Some 23 million of those acres—an area the size of Indiana—are devoted to crops destined for livestock feed.[17] The most frequently irrigated crops are feed corn (some is also used to produce ethanol fuel) and hay, with another 4 to 5 million acres each being planted in soybeans; sorghum, barley, and wheat; and cotton (cottonseed meal is used as livestock feed).

In stark contrast, vegetables, vineyards, and fruit and nut-tree orchards together occupy only 7 million acres of irrigated land.[18] (See figure 3.)

The amount of water devoted to irrigating alfalfa and other hay—7 trillion gallons annually—exceeds the irrigation needs of all vegetables, berries, and fruit orchards *combined*.[20]

Of the roughly 28 trillion gallons of water used for irrigation each year, about 14 trillion are applied to the grains, oil-

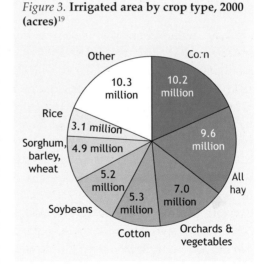

Figure 3. **Irrigated area by crop type, 2000 (acres)**[19]

Other — 10.3 million
Corn — 10.2 million
All hay — 9.6 million
Orchards & vegetables — 7.0 million
Cotton — 5.3 million
Soybeans — 5.2 million
Sorghum, barley, wheat — 4.9 million
Rice — 3.1 million

seeds, pasture, and hay that are fed to livestock in the United States, and an additional 3 trillion gallons are used to produce grains for food or export.[21]

Irrigation Methods Are Often Inefficient

Efficient irrigation methods could help preserve scarce water supplies, but about half of the irrigated acres in the United States use wasteful systems.[22] The least efficient ones either run water down furrows (trenches) or sim-

Level-basin flood irrigation is often used, as on this wheat field, where more-efficient drip irrigation is not appropriate.

ply flood fields. Roughly 45 percent of irrigated acres rely on more efficient systems, such as center-pivot sprinkler irrigation (creating those large circles that can be seen when flying over Nebraska and other Great Plains states).[23] But only 4 percent use highly efficient low-flow systems, such as drip irrigation. Though more expensive than flooding systems, drip irrigation can reduce water use by 30 to 70 percent and increase crop yields by 20 to 90 percent.[24] Adopting better conservation practices and more efficient technologies, which many farmers are now doing, could save tremendous amounts of water.

The timing, as well as the method, of irrigation can waste water and result in "waterlogging, increased soil salinity, erosion, and surface and groundwater quality problems associated with nutrients, pesticides, and pathogens," according to the U.S. Department of Agriculture (USDA).[25] In 2003, only 8 percent of farmers who irrigated their crops measured the moisture content of their plants or soil before irrigating.[26] University of California at Berkeley researchers found that the use of computer models enabled farmers to use 13 percent less water and increase crop yields by 8 percent.[27]

Irrigation May Be a Bad Investment

Irrigated crops account for about one-half of all crop sales in the United States, even though they are harvested from only one-sixth of all cropland.[28]

Subsidizing—and Wasting—Water

American taxpayers provide lavish funding for water projects, mostly benefiting large-scale agriculture and meat-eating consumers. In 1988, the Congressional Budget Office estimated that from 1902—when federal irrigation projects began—through the 1980s, federal subsidies totaled between $34 billion and $70 billion.[29] The World Resources Institute estimates that the federal government—taxpayers—pays an average of 83 percent of the costs of irrigation projects.[30]

Taxpayers help farmers in two ways. First, tax dollars are used to build the systems, then farmers buy water from the projects at a fraction of the cost of pumping or diverting the water. For example, the actual cost of water from the Central Arizona Project, which in 1993 began diverting water for irrigation from the Colorado River, is $209 per acre-foot—yet farmers in Arizona pay only $2 per acre-foot, according to the Congressional Budget Office.[31] Similarly, the full cost of delivering water from the Central Utah Project is $400 per acre-foot, but farmers pay only $8 per acre-foot.[32] In a 2004 study of California water subsidies, the nonprofit Environmental Working Group (EWG) found that American taxpayers are providing up to $416 million per year for California's Central Valley Project. On average, farmers in the Central Valley pay about $17 per acre-foot of water. In stark contrast, Los Angelenos pay about $925 per acre-foot for the water they use. Of the 6,800-plus farms in the Central Valley Project, the top 341 largest were given access to about half of the subsidized irrigation water.[33] Those large farms have little incentive to use the cheap irrigation water efficiently. According to EWG, California's Central Valley has long suffered a host of environmental problems due to over-irrigation, including "devastation of fish and wildlife habitat and severe toxic pollution."

Using irrigation to increase yields means that less land is required to meet the same production goals (it also may contribute to over-production).

In the case of feed crops, the USDA estimates that 100 gallons of irrigation water generates only a few cents in increased farm revenue—hardly a great bargain.[34] The same water could be used for more lucrative purposes. For example, an irrigated acre of corn yields about 163 bushels, which in 2002 was worth about $383. In contrast, 1 irrigated acre could produce about $2,400 worth of potatoes or $4,100 worth of apples.[35] The nonprofit Natural Resources Defense Council estimated that "a 60-acre alfalfa farm using 240 acre-feet of water would generate approximately $60,000 in sales. In contrast, a semiconductor plant using the same amount of water would generate 5,000 times as much, or $300 million."[36,*] While a 60-acre farm could employ as few as 2 workers, the semiconductor plant would

---

*An acre-foot is the amount of water it takes to cover 1 acre of land to a depth of 1 foot.

employ about 2,000. In an analysis of water needs in Western states, the Congressional Budget Office concluded that scarce water supplies should be reallocated from agricultural practices to more economically productive uses to improve what it termed "net social welfare."[37]

Livestock's Consumption of Water Is Huge—and Growing

Farm animals directly consume about 2.3 billion gallons of water per day, or over 800 billion gallons per year. Another 200 billion gallons are used to cool the animals and wash down their facilities, bringing the total to about 1 trillion gallons.[38] That is twice as much water as is used by the 9 million people in the New York City area.[39] Although water use for livestock accounts for a tiny share of national water consumption—about 0.5 per-

Cattle on this treeless, pondless California feedlot need a lot of water to beat the heat.

cent—it is the fastest-growing portion, both in terms of water to drink and the "virtual" water used to grow grains, oilseeds, hay, and pasture.[40] From 1990 to 1995, most categories of water (surface and ground) consumption fell, but water for public use* grew by 4 percent and water use for livestock (including fish farming) grew by 13 percent.[41] Combined with the growing number of livestock over the past 20 years, the increasing number of large cattle feedlots and industrial hog farms may contribute to the rising demand for water.[42] Hog farms use large volumes of water to prepare manure for storage in huge lagoons (see "Manure Lagoons: Accidents Waiting to Happen," p. 94), and feedlots employ misting systems to cool cattle. On traditional farms, in contrast, livestock might find shade or other natural ways to cool off.

---

*Public use includes water withdrawn by public or private water suppliers to use for home, commercial, industrial, or municipal (for example, firefighting and street cleaning) purposes.

Manure Lagoons: Accidents Waiting to Happen

Manure lagoons are supposed to provide safe storage. One maker of lagoon liners advertises "long-term durability, resistance to weathering and low maintenance...can withstand normal environmental exposure for well over 30 years."[43] But sometimes accidents happen. Then, tidal waves of foul-smelling, bacteria-laden liquefied manure flood the land and pollute the water. Just such an environmental disaster happened in June 1995 when an 8-acre cesspool breached (due partly to an unauthorized alteration) and spilled 22 million gallons of waste from the Oceanview Hog Farm into North Carolina's New River Basin. That was the state's largest-ever spill. The waste poured onto nearby farmland, made its way into the river, and robbed the water of much of its oxygen. Thousands of fish were killed, and 364,000 acres of coastal wetlands were closed to shellfishing.[44]

Upstate New York experienced the same kind of manure accident in August 2005 when, according to the Associated Press, "an earthen wall blew out, sending the liquid into a drainage ditch and then into the [Black] River." The "liquid" was 3 million gallons of dairy cow waste—a fish-killing "toxic tide" that was predicted to reach Lake Ontario several days later.[45]

Modern Farming Practices Pollute Water

Irrigation water, pesticides, fertilizer, manure, drugs...they are all widely used or produced on farms, and they often end up polluting nearby streams. The Environmental Protection Agency (EPA) has estimated that "agriculture generates pollutants that degrade aquatic life or interfere with public use of 173,629 river miles (i.e., 25% of all river miles surveyed) and contributes to 70% of all water quality problems identified in rivers and streams."[46]

The pollution, if great enough, kills fish and other aquatic life, prevents people from swimming, reduces crop yields, and impairs drinking water.

Irrigation Leads to Erosion, Runoff, and Salinization

In addition to wasting water, irrigation can degrade the environment. Erosion affects over 20 percent of America's irrigated cropland. When furrows are used to channel irrigation water, sediment runoff often exceeds 9 tons—and sometimes even reaches 45 tons—per acre. Center-pivot sprinkler irrigation causes soil losses as high as 15 tons per acre. The financial cost of replacing nutrients from lost soil runs into billions of dollars annually (see "Erosion," p. 76).[47] In southern Idaho, for example, irrigation-induced erosion has reduced overall crop-yield potential (the estimated seasonal maximum yield) by about 25 percent.[48]

Eroded soil pollutes waterways. The USDA considers sediment from eroded soil to be the "largest contaminant of surface water by weight and volume."[49] In addition, excess irrigation water may pick up contaminants and carry them to rivers and streams. Those contaminants commonly include pesticides and heavy metals (which can contaminate fish) and nutrients from manure or fertilizer (which can lead to algal blooms and loss of oxygen).[50] In California, selenium—which is a naturally occurring element in soil—was so highly concentrated in irrigation water runoff that it caused an epidemic of deformities in migrating waterfowl, including hatchlings born with no eyes or feet (see photo).[51]

These sibling stilt embryos show the effect of selenium contamination. The embryo on the right came from an egg with relatively low selenium content and is normal in outward appearance for this incubation stage. The embryo on the left came from an egg with highly elevated selenium content and exhibits overall stunting (compare the legs of the two embryos), lacks eyes, and has a malformed right foot.

Water extracted from lakes and streams may contain pollutants, such as long-banned pesticides. When that water is applied to farmland, some of it evaporates, leaving behind higher concentrations of those pollutants. In other cases, pollutants settle at the bottoms of streams and lakes, causing them to concentrate and degrade water quality.[52]

Perhaps the most serious danger posed by irrigation to agriculture and the environment is salinization. Water—especially surface water—naturally contains salts. Irrigation water carries those salts onto cropland. When the water evaporates, salts are left behind. Salt buildup can reduce crop yields, and, in extreme cases, may force farmers to abandon once-fertile land. Most estimates put the affected acreage at about 10 million acres, or almost 20 percent of all irrigated land.[53]

Fertilizers, Including Manure, Suffocate Water Life

Fertilizer is a critical contributor to modern agriculture's extraordinary productivity. The fertilizer industry suggests that if farmers stopped using fertilizers, yields of some crops would drop by 30 to 50 percent.[54] However, the heavy use of fertilizers impairs water quality and harms aquatic life.

About half of the 21 million tons of fertilizer used annually in the United States helps produce feed for America's livestock (additional fertilizer is used to grow feed that is exported).[55] Corn, wheat, and soybeans—all major animal-feed crops—are the first-, second-, and fourth-leading consumers of fertilizer, respectively.[56] Farmers treat cornfields with some 232 pounds of fertilizer per acre.

Fertilizer runoff into U.S. waterways is steadily increasing. The industry's Potash and Phosphate Institute estimates that before North America was settled by Europeans, nitrogen runoff into the Mississippi River Basin was 0.7 to 2.1 pounds per acre per year.[57] Sediment studies found protozoa that lived in the area from 1700 until 1900, but could not survive in low-oxygen waters thereafter.[58] That suggests that hypoxia was not a problem until farmers began applying large amounts of fertilizers. The U.S. Geological Survey (USGS) estimates that the average level of nitrogen runoff is now 4 pounds per acre per year, with some areas discharging as much as 50 to 100 pounds.[59] The concentration of dissolved nitrogen (and phosphorus) in the Mississippi River has doubled over the past century, and each year that enormous river discharges 1.8 million tons of nitrogen into the Gulf of Mexico.[60]

According to the EPA, runoff from fertilizer and manure is the biggest polluter of lakes and ponds and among the top five polluters of rivers and streams.[61] When those nutrients wash into waterways, they promote excessive growth of aquatic plants and algae. That increased growth leads to oxygen depletion and eutrophication, which occurs when the decomposition of vegetation absorbs almost all of the available oxygen in the water (hypoxia). Aquatic species then either suffocate or, if they can swim, are forced out of the affected area. As Drew Edmondson, attorney general of

Phosphate Mines Despoil Land, Air, and Water

Before phosphate can be used as fertilizer for feed grains and other crops, it must be mined. Phosphate is strip-mined from near-surface deposits in Florida and Idaho and turned into fertilizer, leaving rivers polluted and landscapes dotted with 200-foot-high hills of slightly radioactive phospho-gypsum by-products.[62] In Idaho, phosphate deposits are located within the greater Yellowstone ecosystem, so mining there threatens the integrity of one of America's most treasured national parks. Indeed, two phosphate refineries in Idaho and one in Florida have been condemned as Superfund sites, ranking them among the nation's most contaminated spots.[63]

Phosphate rock typically is contaminated with heavy metals that are released during the mining process.[64] In Idaho, runoff from phosphate mining has polluted nearby soil and streams with selenium. On one occasion, over 500 sheep died from grazing on heavy-metal-laden grasses near mines, and signs by streams near mining sites warn that the fish may be unsafe to eat.

Phosphate fertilizers—12 million tons of which are produced annually—are made by treating phosphate rock with strong acids.[65] Producing 1 ton of phosphate takes almost 3 tons of sulfuric or phosphoric acid.[66] Those highly corrosive chemicals cause both air and water pollution. One such pollutant is hydrogen fluoride, deemed hazardous under the 1990 Clean Air Act.[67] Chronic exposure to hydrogen fluoride weakens the skeleton, and high concentrations can irreparably damage any tissue in the body. Many phosphate factories also produce phosphoric acid, some of which escapes into the air, where it hovers as a mist that irritates mucous membranes in the eyes, nose, and throat.[68]

Oklahoma, put it when he sued Tyson Foods and 13 other Arkansas poultry companies for polluting local waters, "It's nice to have green land. It's not so nice to have green rivers."[69]

In 1974, scientists discovered that bottom-dwelling aquatic life could not survive in parts of the Gulf of Mexico during the summer. In 1985, that "dead zone"—which emerges each summer—covered about 3,100 square miles. By 1999, the dead zone had doubled in area, and in 2002 it measured 8,500 square miles.[70] That represents an area the size of New Jersey in which aquatic life—including such commercially valuable species as the brown shrimp—cannot survive.[71] Shellfish, starfish, sea anemones, and most other slow-moving animals died off 30 to 40 years ago, leaving the area to a few species of worms.[72]

The dead zone is caused largely by agricultural fertilizer runoff from Midwestern farms that ends up first in the Mississippi River and then the Gulf. Nutrients from agriculture—two-thirds from fertilizer and one-third from manure[73]—account for 80 percent of the nutrient loading in the Mississippi.

Reducing nitrogen losses from agriculture would be the most cost-effective way to reduce hypoxia in the Gulf of Mexico. The National Science and Technology Council, which coordinates the federal government's science policy, estimated the cost of reducing nitrogen runoff from

agriculture at 40 cents for each pound of nitrogen kept out of the Gulf. In contrast, reducing the nitrogen flows from industrial and municipal "point" (that is, definitively identifiable) sources would cost $5 to $50 per pound of nitrogen removed.[74]

In December 2004, Stanford University researchers provided new evidence linking fertilizer runoff to "massive blooms of marine algae in another region."[75]

This summertime satellite photo of the Gulf of Mexico shows where decomposition of phytoplankton that had been fed by fertilizer created an oxygen-poor environment hostile to marine life—the "dead zone." Reds and oranges indicate the most affected areas.

They used satellite imagery to study Mexico's Yaqui River Valley—one of that country's most highly farmed areas. The valley is fertilized and irrigated in cycles over a six-month period, with waters draining into the Sea of Cortez—a long stretch of ocean that separates the bulk of Mexico from the peninsula of Baja California. The researchers saw algal blooms covering up to 223 square miles of the sea. Those blooms appeared after each irrigation cycle, suggesting that fertilizer from irrigation runoff was the culprit.

Manure Contaminates Water, but No Treatment Is Required

Before entering waterways, water polluted with human or other waste is processed in accordance with EPA regulations, which set strict limits on contaminants. This water—from pipes, ditches, and other easily identifiable sites—must be treated and purified, usually at a municipal water treatment plant.[76]

In contrast, livestock manure is *not* regulated by any standards analogous to those that control human waste, and farmers are not required to treat it. Rainwater frequently carries manure downhill from pastures and feedlots into waterways, and some manure leaches into the soil. The EPA recently began to ameliorate the problem by requiring the largest factory farms to obtain permits under the National Pollution Discharge and Elimination System rule—the same rule that governs major industrial and municipal polluters. However, only the largest concentrated animal feeding operations (CAFOs) with 1,000 or more cattle, 2,500 or more hogs, or 30,000 or more broiler chickens are covered by the new rules. The EPA has estimated that the new requirements will reduce nitrogen releases by 110 million pounds and phosphorus releases by 56 million pounds—about a 25 percent reduction in each.[77] Although that is a good start, it still means that, at most, 20,000 of the more than 450,000 CAFOs in the country will have to obtain permits.[78] The remainder will continue to handle excess manure by storing it in lagoons or holding tanks, or by spraying it on fields—all methods that fail to protect public health and the environment adequately.

Where There's Manure, There's Ammonia

At concentrations greater than 2 milligrams per liter (mg/l) of water, ammonia can kill aquatic life.[79] Untreated human sewage has an ammonia concentration of about 50 mg/l. Wastewater treatment plants must limit ammonia in effluent to 4 mg/l in the winter and 1.5 mg/l in the summer. Yet concentrations of ammonia in raw livestock manure can exceed 10,000 mg/l. Concentrations in streams in rural Illinois, for example, range from 26 mg/l to 1,519 mg/l. Between 1985 and 1990, the Illinois Environmental Protection Agency attributed 58 different fish kills—some of which destroyed entire fish populations—to pollution from livestock wastes, though whether ammonia was the primary cause is uncertain.

Ammonia releases from the growing number of factory farms are affecting more and more watersheds. Expanded poultry production in Delaware has increased ammonia releases by 60 percent. Delaware water feeds into the Chesapeake Bay, which receives 81 percent of its ammonia from livestock releases. In North Carolina, ammonia releases have doubled over the past 20 years as hog production tripled.[80]

Ammonia (in the ionized form of ammonium) may be deposited into waterways as it floats back to the Earth's surface or is carried down in rainfall. Ammonium contributes primarily to air pollution, but also can acidify water and increase algal blooms and eutrophication.[81]

Using too much manure on cropland may pollute waterways and soil with dangerous bacteria and excess nutrients. In the upper Midwest, 20 feet of soil protect the water table, reducing the risk that contaminants will reach that water. However, in large areas of North Carolina, the water table lies just 3 feet below the ground, dramatically increasing the chances of contamination.[82]

Pesticides Wash Off of Farmland

The USDA estimates that 5 percent of agricultural pesticides are washed away from farmland through runoff, erosion, and leaching.[83] That threatens the safety of drinking water in many farming regions, where groundwater supplies up to 95 percent of the water used for domestic purposes.[84] In California's heavily farmed San Joaquin-Tulare Basin, at least one pesticide was found in 59 of 100 samples taken from groundwater wells.[85] A 1998 USGS study found the herbicide atrazine in 38 percent of groundwater samples tested; groundwater is the source of most drinking water. Metolachlor was found in 14 percent of groundwater samples.[86] The pesticides only occasionally exceeded drinking water standards, but because the USGS found so many (39) different pesticides—the majority associated with livestock feed production—the cumulative effects of several pesticides acting together might be causing unexpected kinds of harm. Moreover, for several decades, pesticides have been accumulating in bodies of water larger than those tested by the USGS. For example, Lake Superior now contains almost 80,000 pounds of atrazine. In 1991, over 540,000 pounds of atrazine washed down the Mississippi River.[87] Glyphosate, another widely used herbicide, has been detected in about a third of all streams in the Midwest. Its degradation product—aminomethylphosphonic acid—has been found in almost 70 percent of those streams.[88]

Antibiotics in Manure Contaminate Water

In 2002, the USGS found low levels of 22 different antibiotics in a national survey of organic chemical contamination in 139 streams.[89] Those crucial medicines were the eighth-most commonly detected family of chemicals in the survey (about the same as insecticides). The USGS study did not determine the sources of the antibiotics, but presumably those found downstream of livestock operations came mostly from agricultural uses, while those found in urban areas came largely from human uses. The presence of antibiotics in rural streams reflects the mountains of antibiotic-laden manure produced each year and suggests that those antibiotics could lead to resistance among all sorts of bacteria. It's unclear if that poses any risk to humans or wildlife, but prudence would indicate the value of minimizing the drugs' presence (see "Factory Farming's Antibiotic Crutch," p. 68).

What It All Means

Extracting water for irrigation and livestock use is one of many areas in which agriculture is exceeding the limits of sustainability and harming the environment. In some parts of the country, groundwater supplies are being gradually but inexorably and irreplaceably depleted. The ecological damage from extensive and excessive irrigation includes soil erosion, fish and bird poisonings, impaired fish habitats (threatening the very survival of Coho and Chinook salmon throughout much of the Pacific Northwest), and damage to roads and houses as the land below them sinks—mostly to raise crops that generate only pennies for every 100 gallons of irrigation water. In addition, the fertilizer and pesticides used to grow feed grains and other crops, and the manure from the animals that eat the feed, pollute water all the way from the farm to the nation's great rivers and the oceans.

Reducing the number of animals raised for food and raising cattle on rangeland instead of in feedlots are obvious ways to reduce water consumption in the West and Great Plains. A complementary approach is to use water in more sustainable and productive ways. Cutting back on meat consumption would protect waterways from pollution caused by fertilizer production, runoff from chemical fertilizer and manure, and soil erosion. Of course, producing more fruits, vegetables, beans, and nuts still would require water, but far less than is needed to produce animal products.

Argument #5.
Cleaner Air

In February 2001, two workers at a California dairy farm were ordered by their foreman to climb into a manure storage pit to unclog a drainpipe. Soon after descending into the 30-foot-deep pit, José Alatorre—standing in manure up to his knees—began to complain that the air quality was poor. Moments later, he attempted to climb out of the pit, but was overcome by the noxious gases given off by the manure. Before losing consciousness, he called out for help. When co-worker Enrique Araisa climbed down to help Alatorre, he too succumbed to the gas. Both men drowned in the putrid, liquefied waste.[1]

Half a century ago, farms typically raised dozens—or at most, hundreds—of chickens, pigs, or cattle in their barnyards and on their pastures. Today's production facilities—it's hard to use the word "farm"—are so large and house such huge numbers of densely packed livestock that they would have been inconceivable to farmers half a century ago. Consider: While

our population almost doubled between 1950 and 2003, the amount of farmland fell by 22 percent to 939 million acres, and the number of farms plummeted by 63 percent to 2.1 million.[2] At the same time, however, red meat production more than doubled to 47 billion pounds per year, and chicken production rocketed more than 20-fold to 41 billion pounds annually.[3,*]

In short, far more animals are being raised on far fewer farms. One result is massive environmental harm, including air pollution.

- Manure and urine on factory farms release foul-smelling gases that can sicken humans and animals and harm the environment.

- Odors from large-scale livestock operations can cause drowsiness, headaches, and poor concentration in nearby residents.

- In 2000, methane belched out by cattle and generated by livestock manure had the same impact on global warming as the carbon dioxide produced by about 33 million automobiles.

Whereas problems from livestock and manure odors used to be relatively rare, today's high density of animals means that feces and urine from vast herds and flocks stink up the air, afflicting anyone unfortunate enough to live or work downwind.

Livestock excreta—including that stored in foul-smelling manure "lagoons" larger than football fields—is only the most obvious form of air pollution due to animal agriculture. The production and use of fertilizer to nourish feed grains release toxic substances that despoil the atmosphere, dust carries germs and risky chemicals, pesticides are blown far and wide, cattle belch up great volumes of a greenhouse gas, and even milling grain to make animal feed generates clouds of dust.

Factory Farms Emit Noxious Gases

Learning about the various air pollutants produced by today's farms is almost like taking a chemistry lesson. From the most harmful, ammonia, to the most offensive, odor, a toxic cornucopia of chemicals harms everything from human lungs to the Earth's atmosphere (see figure 1).

Ammonia

Livestock are the largest source of ammonia releases on Earth. In the United States, animal agriculture—especially from manure and fertilizer—accounts for about 82 percent of ammonia releases.[4] Cattle waste is

---

*The weight of meat or eggs produced, rather than the number of animals raised, is our growth gauge, because not only are more animals being raised, but breeds of livestock generally have gotten bigger. Data for chicken are for 1950 and 2002.

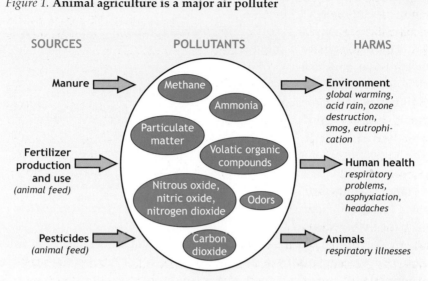

Figure 1. **Animal agriculture is a major air polluter**

responsible for 43 percent of that discharge, swine 11 percent, and poultry 27 percent. Most of the ammonia comes from feces, but urine adds to the burden.

Applying manure to farmland allows large amounts of ammonia to evaporate into the air.[5] There, the ammonia reacts with sulfur- and nitrogen-containing gases. Those gases can cause respiratory and other health problems, as well as contribute to smog and acid rain.[6]

Ammonia irritates mucous membranes in humans at concentrations of about 10 parts per million (ppm).[7] The National Institute for Occupational Safety and Health recommends a maximum safe exposure of 25 ppm. While a well-ventilated hog shed has concentrations of 10 to 20 ppm, sheds tend to be poorly ventilated during the winter, and ammonia levels can reach 100 to 200 ppm. At those concentrations, farmworkers are likely to suffer intense irritation of the skin, eyes, nose, throat, or lungs. The hogs, which breathe the polluted air continuously, have an increased risk of pneumonia and other respiratory illnesses.

The manure lagoons on industrialized hog farms release large amounts of ammonia into the air.[8] A study by the U.S. Department of Agriculture's (USDA's) James Zahn, an expert on factory-farm emissions, found that a 2-acre swine manure pond produced more than 100 pounds of ammonia per day on over 200 days in a single year. On one hot day, the Missouri lagoon under study released 277 pounds of ammonia.[9]

British researchers studied plant life near a complex that housed 350,000 chickens. They blamed the invisible cloud of ammonia for eliminating half of the plant species found near the chicken sheds. The number of species increased with the distance from the livestock buildings. However, the trees, grasses,

Lagoons on hog farms, such as this one in Iowa, may use hillside terraces to purify wastewater, but they still emit ammonia and other air pollutants.

and mosses that survived had high concentrations of nitrogen in their tissue. At four-tenths of a mile from the poultry houses—the farthest the scientists examined—the nitrogen content of plants was twice the normal level.[10]

Because ammonia is highly water soluble, rain deposits airborne ammonia onto land and into waterways. Once there, it can increase the acidity of soil and water, decrease the productivity of forests and coastal waters, and disrupt ecosystem biodiversity.[11] Ammonia from chicken houses has been deemed a "silent killer of the Chesapeake Bay," the nation's

The Effects of Air Pollution: Clouded in Uncertainty

Despite uncertainty over the exact amount of damage done by air pollutants generated by livestock, those compounds clearly harm humans, animals, and the environment. The National Research Council identified ammonia and odor as "major" concerns. Methane, nitrous oxide, nitric oxide, hydrogen sulfide, and particulate matter are "significant" concerns.[12]

Most of the research on the health effects of the gases emitted by feedlots and other large, concentrated livestock operations has focused on brief exposures to high concentrations. But residents living near factory farms are chronically exposed to lower concentrations, the effects of which are harder to study. And the health effects of certain types of emissions—most notably odor—have only begun to receive serious attention.

Synergistic effects from individual pollutants may exacerbate the damage. For example, chemical reactions between carbon monoxide and other pollutants—nitrogen oxides and volatile organic compounds—produce ground-level ozone (smog),[13] which can cause asthma, bronchitis, emphysema, and other illnesses.

largest—and once probably richest—estuary.[14] Over the past 30 years, the bay has been severely polluted by ammonia and other gases that evaporate from manure at nearby chicken farms. On the Delmarva Peninsula, which stretches along the eastern side of the bay, the leavings of almost 600 million chickens grown on 2,100 farms release some 20,000 tons of ammonia each year. In the summer of 2004, 27 percent of the nitrogen deposited into the bay came from ammonia that had risen from surrounding farms into the atmosphere and then drifted down into the water. Once in the water, the ammonia contributes to algal blooms that deprive waterways—and their aquatic life—of oxygen, a process called eutrophication.[15]

When asked about the odor around the bay, a local soybean farmer lamented, "When the winds change, [the smell] can get so bad outside you got to close the house up with all the windows shut."[16] The *Chicago Tribune* observed that the "stench and noxious gases from large-scale livestock farms…are tearing apart some rural communities."[17]

Methane

Livestock—primarily cattle—generate methane, a greenhouse gas, when they digest food and when bacteria digest manure. Cattle's belching and flatulence are responsible for 19 percent of all methane gas released in the United States.[18] Another 13 percent is released by anaerobic bacteria, which thrive in the almost oxygen-free manure lagoons located mostly on hog farms.

At concentrations of 5 to 15 percent, odorless methane can asphyxiate people.[19] It causes occasional deaths across the United States, mostly among farmworkers—such as the two mentioned at the beginning of this chapter who were cleaning manure storage tanks.

Methane traps heat in the atmosphere—a process that is slowly raising the Earth's temperature and causing profound climatic and environmental changes. On a pound-for-pound basis, methane is 23 times more conducive to global warming than carbon dioxide.[20] In 2000, livestock and manure lagoons released an amount of methane that was equivalent in environmental damage to the carbon dioxide from about 33 million automobiles.[21]

Nitrous Oxide

Nitrous oxide is a greenhouse gas that is about 300 times more powerful than carbon dioxide.[22] About 25 percent of the nitrous oxide from animal agriculture in the United States comes from bacteria that digest animal waste.[23] Nitrous oxide is produced by anaerobic soil bacteria when manure or fertilizer is applied to land. Cattle waste accounts for over 90 percent of the nitrous oxide derived from livestock manure.[24]

The only human-generated source of nitrous oxide larger than animal waste is fertilizer applied to cropland. Because more fertilizer is used for growing livestock feed than anything else, raising animals for meat and dairy foods is the main driver of the two biggest sources of nitrous oxide from human activities in the United States. Nitrous oxide accounts for 6 percent of the greenhouse effect in the lower atmosphere.[25] When it migrates to the upper atmosphere, nitrous oxide catalyzes ozone-destroying reactions.[26]

Nitric Oxide and Nitrogen Dioxide

Another nitrogen-based air pollutant is nitric oxide. It comes mainly from the burning of fossil fuels but is also produced when bacteria in the soil digest nitrogen compounds. The nitrogen from livestock waste, cropland, and fertilizer "feeds" those bacteria, accounting for 5 percent of the nitric oxide generated by human activity. In Illinois, for example, over one-quarter of the nitric oxide released comes from the many cornfields.[27] Farm equipment also contributes to emissions through fossil fuel combustion.[28]

Sunlight converts nitric oxide into nitrogen dioxide. Those two compounds, which are referred to collectively as nitrogen oxides, or NO_X, can degrade the environment in several ways, including increasing

Making Fertilizer, Making Pollution

Natural gas is made into ammonia, which is then used directly as a fertilizer or used to produce urea and ammonium nitrate fertilizers. Nitrogen fertilizer factories discharge ammonia and nitric acid into the air.[29] They also release carbon monoxide, a greenhouse gas; fine particulate matter that can clog capillaries in the lungs and cause respiratory infections; sulfur dioxide, which readily converts into sulfuric acid and contributes to acid rain; and nitrogen compounds that contribute to acid rain, global warming, and ozone depletion.[30] Worldwide, fertilizer production generates 1 percent of all greenhouse gases.[31]

The lower ozone levels expose humans to higher levels of ultraviolet rays. Meanwhile, the acid rain degrades for-

> *Conversion Fact*
>
> The amount of energy used annually to produce the 22 billion pounds of fertilizer used to grow animal feed in the United States could support roughly 1 million people for one year.[32]

ests, lakes, and streams. The gases that cause acid rain also form fine sulfate and nitrate particles that increase the risk of heart and lung disorders, including asthma and bronchitis.

ozone levels in the lower atmosphere. According to Vaclav Smil, a global-ecosystems expert at the University of Manitoba, ozone "impairs lung function, injures cells, limits the capacity for work and exercise, and lowers the resistance to bacterial infections."[33] NO_x also can form nitric acid or increase airborne particulate matter, contributing to both smog and acid rain. Once deposited onto land, NO_x increases the acidity of soil and decreases biodiversity, including of plant life.[34] Deposited in water, NO_x increases the acidity and promotes eutrophication.

Hydrogen Sulfide

Hydrogen sulfide, another invisible gas released by intensive animal agriculture, is the gas with the distinctive "rotten egg" smell. It is produced by anaerobic bacteria in animal manure stored under moist conditions. Liquefying the waste—as is often done on factory farms—exacerbates the problem.[35]

Nationally, the amount of hydrogen sulfide generated from livestock manure is small, but at the local level, the gas can be a serious problem. Even a concentration as low as 2 ppm can cause headaches. Slightly higher levels can cause respiratory, cardiovascular, and metabolic problems. When swine waste is agitated—which occurs when storage tanks are drained—hydrogen sulfide concentrations near the tanks can reach 200 to 1,500 ppm and seriously harm human health.[36]

As with methane, hydrogen sulfide vapors have killed farmworkers in and around manure storage tanks. The National Institute for Occupational Safety and Health recommends that exposure levels be kept below 10 ppm and that individuals evacuate if levels exceed 50 ppm.

Toxicity of Hydrogen Sulfide[37]

- 2 ppm: headaches
- 2-10 ppm: respiratory, cardiovascular, and metabolic problems
- 50 100 ppm: vomiting and diarrhea
- 200 ppm: immunological problems
- 500 ppm: loss of consciousness
- 600 ppm: often fatal

An Ohio man suffered memory losses, poor balance, a stutter, and other symptoms that his doctor blamed on a large hog farm half a mile from his house.[38] The doctor pinpointed high levels of hydrogen sulfide as the culprit, but other gases also could have been involved. "If I could sell the house, I would move in a second, but I don't know where to go," the man told the *New York Times*.

Hydrogen sulfide also harms animals. Factory farms commonly use slatted concrete floors to drain manure into storage tanks directly below the animals. That practice is most common on hog farms, but is also sometimes used with cattle. Spending their lives above a pit full of liquefied manure continuously exposes the animals to hydrogen sulfide and other harmful gases. Hogs exposed continuously to just 20 ppm of hydrogen sulfide become anxious and afraid of light.[39] Animals have died when they breathed the higher levels of hydrogen sulfide that occur when waste is agitated.[40]

Volatile Organic Compounds

A broad array of volatile organic compounds (VOCs) form and then pollute the air when manure breaks down. Those chemicals have a carbon back-bone, which is coupled with hydrogen, oxygen, fluorine, chlorine, bromine, sulfur, or nitrogen. VOCs from factory farms include organic sulfides, alde-hydes, amines, and fatty acids.[41]

VOCs may irritate the skin, eyes, nose, and throat. They can be trans-ported by nerve cells directly to the brain, thus affecting the central nervous system. VOCs absorbed by the lungs, digestive tract, and skin can affect metabolic and physiological processes. If inhaled, VOCs can increase the risk of respiratory infections, such as pneumonia, and might weaken the overall immune system.[42] In addition, they contribute to the formation of smog and exacerbate the greenhouse effect. Regulatory agencies have not yet set exposure limits.[43]

Odor

Odor is the most readily perceived environmental problem caused by large-scale animal farming. Although odor is downplayed by some economists as only a minor nuisance that might reduce neighbors' property values,[44] it may have serious health consequences that we are only now beginning to understand.

Livestock operations generate a cafeteria of odoriferous chemicals, including ammonia, hydrogen sulfide, and VOCs. One study found 331 distinct odor-causing compounds in hog manure.[45]

Odors from factory farms irritate the eyes, nose, and throat. They also cause headaches, drowsiness, allergic reactions, breathing difficul-ties, and higher incidences of diarrhea. In one study, stench—euphemis-tically termed "malodor"—was associated with an immunosuppressive effect that increases the risk of disease and infection in both humans and animals.[46]

Besides causing physical problems, odors have a profound effect on mood and performance. One study found that "persons living near…in-tensive swine operations who experienced the odors had significantly more tension, more depression, more anger, less vigor, more fatigue, and more confusion" than people living farther away.[47]

Football field-sized mountains of cattle manure stink up the neighborhood and endanger nearby streams.

Particulate Matter

Intensive animal agriculture generates immense amounts of "particulate matter" that comes primarily from animal hair, dried manure, and dander (small flakes of skin, feathers, or hair). The fine dust is easily scattered by wind and animal movement. The problems it causes are most severe around cattle feedlots because the ground—unlike pasture—is bare and exposed to the wind.

The health effects of particulate matter depend, in part, on its size. Particulate matter typically is divided into two categories: PM2.5, which includes all particles smaller than 2.5 microns;* and PM10, which includes everything smaller than 10 microns. Both categories cause environmental and health problems, but PM2.5 is a greater threat because the particles' small size allows them to penetrate even the tiniest airways in the lungs and cause respiratory illness and infection.[48] Moreover, the dander in particulate matter causes some people to develop asthma or allergies to cattle, hogs, or sheep.

Particulate matter produced on farms may carry viruses, bacteria, and fungi, as well as traces of the antibiotics added to animal feed. One study of the air in a large pig-feeding operation found bacteria, some of which were resistant to several antibiotics typically given to hogs.[49] Whether one could become infected upon breathing the air depends on the concentration of the bacteria. In addition, the antibiotics themselves have been discovered in the air and could conceivably cause allergic reactions.[50]

On the environmental front, the particulate matter sent airborne from feedlots and farms can react with ozone, generating the low-hanging clouds

---

*A micron is one-thousandth of a millimeter.

Pesticides in the Air

Farmers intend for their pesticides to do their handiwork on crops or soil, but when the chemicals are sprayed, some amount inevitably drifts away with air currents. Also, pesticides may volatilize (evaporate) from the field. Through those two processes, as much as 40 to 60 percent of the pesticides applied to crops may reach the Earth's atmosphere.[51]

The pesticides eventually come back to Earth—primarily in rainfall—far from where they were applied. Traces of atrazine—the second-most widely used herbicide on feed grains—occurred in 30 percent of the rainfall samples tested in Midwestern and Northeastern states. Metolachlor—the fourth-most commonly used herbicide on feed grains—was found in 13 percent of the samples.[52] About 250,000 pounds of atrazine were deposited by rain into the Mississippi and Ohio River Basins in 1991 alone. Whether the small amounts of pesticides that are blown far from farmers' fields pose any subtle health risks to people or wildlife is not known.

of pollution that once were associated only with urban and industrial areas.[53] As one startling example, California's San Joaquin Valley, home to one-fifth of America's dairy cows, now competes with Los Angeles and Houston for having the most polluted air in the country. A Sierra Club spokesperson told the *Washington Post*, "It's not just a stink that's coming out of these farms. It's a real health threat."[54]

What It All Means

Factory farms produce toxic gases, noxious odors, and particulate matter that make life on the farms and downwind miserable—and unhealthy. The damage from the pollution generated by these operations extends even up to the Earth's atmosphere. Those ills and the welcome trend toward sustainable agriculture notwithstanding, we will never totally return to the less-intensive, less-destructive, but also less-efficient, agricultural practices of yesteryear. While industry and government fight over more protective regulations, one simple step each of us could take is to eat fewer animal products, especially from factory-raised animals. That would reduce the number of livestock and the amount of air pollution they generate.

Argument #6.
Less Animal Suffering

"*Our inhumane treatment of livestock is becoming widespread and more and more barbaric....A Texas beef company, with 22 citations for cruelty to animals, was found chopping the hooves off live cattle....Secret videos from an Iowa pork plant show hogs squealing and kicking as they are being lowered into the boiling water that will soften...the bristles on the hogs and make them easier to skin....Barbaric treatment of helpless, defenseless creatures must not be tolerated even if these animals are being raised for food....Such insensitivity is insidious and can spread and is dangerous. Life must be respected and dealt with humanely in a civilized society.*"
—U.S. Senator Robert Byrd[1]

Many animals die to please our palette. About 140 million cattle, pigs, and sheep are slaughtered annually in the United States—about half an animal for every man, woman, and child (see table 1).[2] Add to that 9 billion chickens and turkeys—30 birds for every American—plus millions of fish, shellfish, and other sea creatures.[3]

The American Meat Institute contends that "Animal handling in meat plants has never been better."[4] That might well be true, but "never been better" falls far short of "good."

There's no easy way to know what constitutes happiness or contentment or pain for a pig, a cow, or a chicken.[5] We can anthropomorphize livestock, imagining how it would feel to undergo some of the same experiences: having our teeth pulled or being castrated without anesthesia, for example. And in many cases, the pain an animal is experiencing is perfectly obvious. However, that approach is considered by some to be too subjective to establish the effects of such practices on animals. New tests are being developed that use the behavioral and biochemical markers of stress to evaluate farm animal welfare. Because the European, but not the American, legal system treats livestock as sentient, conscious creatures, the majority of that research is taking place abroad.

- Industrially farmed chickens are raised in enormous and crowded sheds, may never see the outdoors, and exhibit abnormal behavior. Layer hens live in tiny cages, are debeaked, and are periodically starved to maximize egg production.

- The unnatural high-grain diets of cattle in feedlots sometimes cause liver, hoof, and digestive diseases.

- Pregnant and nursing pigs spend most of their time in pens so small they cannot even turn around in them.

- U.S. farm animals are not legally protected as are laboratory animals.

Food animals are not protected by federal animal welfare laws.[6] In fact, farm animals are specifically exempted from the laws that protect rats, mice, and other laboratory animals. While more than 30 states have livestock anti-cruelty laws, they typically exempt "common" or "customary" practices. Therefore, painful procedures—such as when animals' beaks, horns, tails, or testes are chopped off—are legal because most farmers use them. As Matthew Scully argues in his book *Dominion: The Power of Man, the Suffering of Animals, and the Call to Mercy*, "When the law sets billions of creatures apart from the basic standards elsewhere governing the treatment of animals, when the law denies in effect that they are animals at all, that is not neutrality. That is falsehood, and license for cruelty."[8]

Table 1. **Food animals slaughtered in the United States, 2003**[7]

Animal	Number
Sheep	2,900,000
Ducks	26,000,000
Cattle/calves	33,800,000
Hogs	104,000,000
Turkeys	254,000,000
Chickens	8,900,000,000
Total	9,320,000,000

"Bycatch": Bye Animals

In addition to the land and sea animals intentionally raised or caught for food, millions more die unintentionally as farmers and fishers seek to satisfy our appetites:

- Billions of pounds of commercially useless fish, turtles, and other sea animals are unintentionally caught as "bycatch" and discarded, already dead or dying.
- Wildlife is poisoned by the pesticides applied to crops.
- Farm animals die of injuries or illnesses before they reach the slaughterhouse.
- The egg industry literally shreds millions of male chicks at birth.

In such a lax regulatory environment, agricultural practices that many people consider brutal have become the norm. From birth to death, many animals never see the outdoors. They are caged or otherwise housed in cramped conditions where they sit in their own excrement. That sort of husbandry produces unnatural repetitive behaviors called "stereotypies" that may result in injury to the animals themselves or to nearby animals. Most cattle are fed grain-based diets that may cause ulcers in their stomachs and suffocating gases. Near the end of their short and often miserable lives, livestock are crammed into crowded trucks lacking food and water and transported to slaughterhouses where they sometimes suffer painful deaths.

It is worth recognizing that many seemingly inappropriate or downright inhumane practices have some practical benefits to the animals or the farmers or they wouldn't be done. For instance, indoor confinement of chickens, turkeys, and pigs, while unnatural and sometimes unhealthy, protects the animals from predators, deadly germs such as the avian influenza virus, and harsh weather. The questions are whether those benefits are so great that they outweigh the harm done to the animals and whether alternative methods could reduce animal suffering.

Farm Animals' Unnatural Lives

Separated Early from Their Mothers

The dairy industry obviously has little use for males, so they typically are transferred into veal or beef production systems. Calves are often separated from their mothers within one day of their birth—before they can walk and before they have received from their mothers' milk essential proteins for growth and immunity to germs.[9] The day of separation is traumatic for both mother and offspring, with each bawling for the other.

Early removal from their mothers and subsequent isolation reduce calves' ability to develop normal social behaviors and contribute to the development of abnormal behaviors. Because they are fed from a bucket rather than nursing at teats,

weaned calves miss the opportunity to satisfy an instinctive desire for suckling.[10] That thwarted desire leads calves to lick themselves and other animals obsessively, which results in rumen hairballs. Those hairballs can weigh as much as 8 pounds and occasionally harm the animals.[11] Calves also may try to nurse on each other or induce urination by licking each others' genitalia and then drinking the urine.[12]

Stamped as Property

Beef cattle—especially out West—are often "branded" with a logo indicating their ownership. Branding has been used by ranchers for generations and has deep cultural resonance, if limited utility. Depending on its age at the time of branding, the animal is either pinned on the ground or constrained in a chute. The brand is then impressed into its hide using a blazing hot iron, which creates a third-degree burn; that painful process may be repeated when animals are sold to different owners.[13] Many more humane alternatives for animal identification exist, such as ear tags or retinal imaging, which should consign this outmoded practice to the history books. Furthermore, the threat of mad cow disease highlights the importance of instituting a national system for livestock tracking. Branding is practically useless for that purpose because of its limited information content.

Inconvenient Parts Removed

Castration

Nearly all bulls are castrated, which involves removing their testicles. The most common methods are slitting the scrotum and removing the testicles, blocking the circulation of blood to the scrotal sack with a tight rubber band, breaking the spermatic cord with pliers, or injecting the testicles with an acid or other chemical.[14] All are performed without painkillers.

Most calves are castrated when they are less than a month old. Some argue that young animals feel less pain, but Bernard Rollin, a prominent animal welfare expert at Colorado State University, says there are "no good grounds for believing that pain experience is tied to age. It is well-known that cattle are born precocious, and it would be biologically and evolutionarily incredible that all faculties are formed at birth except pain capacity." In fact, inflicting pain on young animals may lead to chronic pain later in life.[15]

Castration does offer several benefits. It makes steers* more docile, which keeps them from injuring one another in crowded feedlots. It also improves meat tenderness, primarily by increasing the fat content. However, castration is not a unique way to obtain those benefits. Giving cattle more space decreases aggression, too.[16] And tenderness is not an issue with meat from younger bulls and can be improved by aging meat from older bulls.

Cattle that are not castrated have their own virtues. They are more efficient at converting feed to weight gain and therefore reach market weight faster.[17] That means they consume less grain, saving money and natural resources. Cattle ranchers compensate for the slower growth of steers by implanting hormone pellets in their ears to replace those naturally produced by the testicles (see "Sex Hormones on Ranches," p. 131).

Ultimately, it is economics that spurs ranchers to castrate their bull calves. Packers pay less for bulls than for steers, ostensibly because consumers prefer the fattier steer meat. Yet some "boutique" beef producers specialize in bull meat because of a niche demand for its lower fat content. Meatpacking companies, which largely are responsible for determining what price producers will receive, can identify bull and steer carcasses because U.S. Department of Agriculture (USDA) inspectors—despite the absence of any regulatory requirements—prominently stamp carcasses from

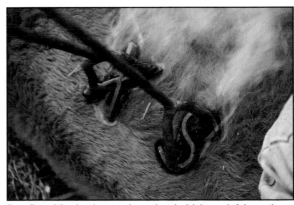

Branding with a hot iron can be replaced with less painful practices.

uncastrated males as "bullock."[18] The bullock stamp essentially punishes ranchers who avoid causing pain to their animals and deliver a leaner, healthier product to consumers.

---

*Bulls are uncastrated cattle; steers are castrated.

Dehorning

The major breeds of dairy cattle grow horns, as do some beef cattle breeds. Horned cattle are still raised because other breeding priorities—rapid weight gain or robust milk production, for example—have trumped the desire to breed the horns out of the cattle.[19] To prevent crowded, stressed animals from injuring each other or their handlers, dairy cattle are dehorned at an early age.[20] The nascent horn is gouged out, cut off, or burned with either a hot iron or chemicals. Although horns are commonly thought of as woody protuberances devoid of sensation, they are actually more similar to teeth—their hard shell covers a rich vascular and nervous network. Dehorning can be extremely painful and may cause extensive bleeding. As with castration, calves typically are not given painkillers when they are dehorned.

Tail Docking

Removing the tails of dairy cattle—another terribly painful procedure—has become increasingly common. Cows' tails often become coated with dirt and excrement, so when they swish their tails to chase off flies, they fling about whatever filth has accumulated on them. Some dairy producers believe that tail swishing increases the risk of mastitis, a painful bacterial infection of the udder, because manure could land on a cow's udder. Another argument for tail docking is that tails may be trampled on by other animals, causing lesions and infections.

Professor Rollin argues that there is "absolutely no scientific basis for claims about the benefits of tail-docking....Removing the tail is another example of attempting to handle a problem of human management by mutilating the animal."[21] Instead of docking tails to prevent mastitis, farmers should clean up dirty stalls. The trampling problem could be avoided by giving the cows more space.[22] All in all, we suspect that American cows would much prefer to live in Sweden, where tail docking is forbidden, local anesthesia or a sedative must be used for dehorning, and cows must be kept on pasture for at least two to four months out of the year.[23]

Cows ♥ Farmers

Lost in the industrial dairy system is the bond between cows and the farmers who care for them. Research has demonstrated that dairy farmers who relate well to their animals get higher yields.[24] Industrial agriculture, however, increases the number of animals per handler, which reduces the interaction between animals and farmers.

Debeaking, Detoeing, and Maceration

Because of the economic losses associated with feather pecking, egg farmers routinely trim off the tips of birds' beaks. Debeaking (see photo) causes

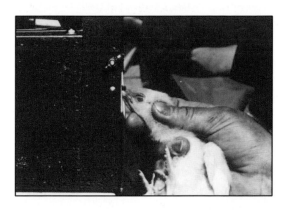

both acute and chronic pain, including pain during eating.[25] To prevent sometimes serious injury during fights, poultry are often detoed.[26]

Treatment of male chicks is even more grotesque. Because the egg industry has no use for those birds, they are summarily killed. The current method of choice is to dispose of the birds in what is effectively a modified wood chipper. Industry parlance describes this as "instant maceration using a specially designed high-speed grinder." Other methods of disposal, considered less humane, include suffocation and crushing.[27]

Confinement in Tight, Unhealthy Quarters

Cattle

Dairy farmers increasingly keep their cows indoors, confined in accordance with industry recommendations of about 20 to 25 square feet per 1,000 pounds of animal.[28] To put that space into perspective, the tiniest car on U.S. roads—the Mini Cooper—occupies about 75 square feet. Its "footprint" would accommodate three adult cows with some room to spare.[29]

Beef cattle are similarly confined during the last several months of their lives, albeit in outdoor feedlots. Those usually give the animals more space than their milked counterparts, but the cattle are limited to only a grassless field of manure instead of pasture.

Dairy cows, once pasture-raised on small farms, increasingly are being raised in confinement on mega-farms.

Pigs

Pigs generally are considered to be the most intelligent of the major live-stock species, which makes their suffering especially inhumane.[30] Unlike beef cattle, which typically are raised on pasture for most of their lives, pigs may spend their entire existence in an individual pen or in a limited space with a small number of other pigs. Gestation crates are used for pregnant sows and farrowing crates for sows that have just given birth. The main difference is that farrowing crates have a side area where the newborn piglets can fit. Those pens usually are only about seven feet by two feet. According to Alberta Pork, a pork producers' association in Canada, "The crate (sometimes called a stall) is a simple pen made of metal that contains the sow in the least possible space that is economically feasible.... Sows housed in a crate

Pregnant pigs are typically held in cramped gestation crates.

cannot turn around, but they can stand up and lie down and take one step forward or backward."[31]

Confined sows suffer health problems not commonly seen in pigs raised outdoors. They have more foot and leg injuries—including fractures—probably as a result of living

in pens with slatted floors.[32] They also have more urinary tract infections, perhaps because the floors on which they lie are dirtied with their own waste. Furthermore, gestation crates increase the likelihood that sows will endure particularly long or painful births; fail to secrete milk; and suffer from "wasting disease," which causes them to gradually lose weight, have a variety of organ problems, and often die. (The bacterial or other cause of wasting disease has not yet been identified.)

Farrowing crates in which sows could give birth to and nurse their piglets were introduced because the sows had a habit of lying on their piglets and crushing them to death. That failure of the maternal instinct is itself partly the result of poor breeding practices. Pigs have been selected for lean meat and rapid growth; somewhere in their breeding history, they lost the ability to protect their young properly. Farrowing crates do help protect piglets, so industrial farm operators argue that tight confinement is a welfare measure. But old-fashioned pigs raised the old-fashioned way normally didn't crush their offspring.

Treating pigs humanely does not necessarily sacrifice productivity. Sweden banned gestation crates in 1994, and the United Kingdom banned them five years later. (Both the European Union and New Zealand are in the process of phasing them out.) In Sweden, pork production actually *rose* after the ban. In Great Britain, pork production fell, but that was due to an ongoing outbreak of post-weaning multisystemic wasting syndrome—an illness that kills young pigs but is not related to the use or absence of gestation crates.[33]

Factory-farmed pigs also must contend with the potentially fatal gases released by their manure. In many operations, manure falls through slats in the floor into a pool directly below the pens. Dangerous gases rise up from that manure. Among them is hydrogen sulfide, which, according to James Barker—a North Carolina State University expert on animal manure nutrients—can produce "fear of light, loss of appetite, [and] nervousness" in pigs.[34] In high concentrations, those fumes can be fatal. Other manure gases, such as ammonia, increase hogs' risk of pneumonia, other respiratory diseases, and convulsions.

Chickens

Layer hens—chickens raised to produce eggs—are housed in stacked rows of tiny "battery cages," typically with five to seven birds per cage. According to an animal welfare organization, a single farm may house up to 800,000 birds at a time.[35] For adult Leghorn chick-  ens, the most widely used breed in the world, academic researchers recommend that each bird be allotted half a square foot.[36] In 2005, the United Egg Producers—a major industry group—increased its recommended allotment from 0.33 square feet per bird to between 0.47 and 0.60 square feet, depending on the size of the hen.[37] That recommendation will be phased in over five years. (The European Union requires 0.5 to 0.6 square feet, and will increase

Concentrated Disasters

The confinement of tens of thousands of chickens and thousands of pigs in small areas is a prescription for mass disaster. When Hurricane Katrina devastated Louisiana, Mississippi, and Alabama in 2005, it was not just people who were affected: Millions of chickens were killed due to power outages and lack of water.[38] The same thing happened in North Carolina in 1999, when the winds and rain of Hurricane Floyd killed more than 2 million chickens and turkeys and hundreds of thousands of hogs.[39]

that requirement to 0.8 square feet by 2012.[40]) Note that an 8½-by-11-inch sheet of paper is 0.65 square feet—about 30 percent larger than the space a hen in the United States is now provided.

Although hens that are less crowded are more productive individually, the poultry industry gets a higher overall yield of eggs by cramming more hens into fewer cages. Rollin, at Colorado State University, notes: "It is nonetheless more economically efficient to put a greater number of birds into each cage.... Though each hen is less productive when crowded, the operation as a whole makes more money with a high stocking density: Chickens are cheap, cages are expensive."[41]

Rollin is also concerned about the wire floors of battery cages, which may injure hens' feet and legs.[42] A chicken may catch its head, neck, or wings in the wire sides of the cage, which could lead to serious injury. Another problem is that the tight confinement does not permit exercise, such as normal wing-flapping and (brief) flying. The absence of exercise increases the incidence of lameness, brittle bones (osteoporosis), and muscle weakness. At slaughter, 6.5 percent of caged hens have broken wings compared to just 0.5 percent of free-range hens. Dust-bathing—another regular activity of chickens and a natural protection against parasites—

also is impossible in the cramped cages.

In contrast to practices in America, Switzerland banned the use of barren cages that lack materials for nesting, and the European Union is in the process of banning them as well.[43]

Broiler chickens, in contrast to layer hens, are raised on sawdust floors in sheds as big as football fields. They are kept together for their entire, albeit brief, six-week lives in groups of 10,000, 20,000, and sometimes even 50,000. Obviously, it is impossible for farmers to monitor the health of individual animals in such a setting. When disease outbreaks occur, they can race through entire flocks and cause widespread death, or, in the case of Exotic Newcastle disease or avian influenza (bird flu), require the slaughter of the entire flock.[44]

The floor covering in a broiler house is not changed during the course of a single flock's life—or even several flocks' lives. Feathers, feces, and feed all become mixed with sawdust. The high acidity of chicken dung that collects on floors can cause burns on chickens' feet and legs.[45]

Pushed to Produce

Dairy Cattle

As the rate of milk production has risen—it is now six times as high as 100 years ago[46]—dairy cows increasingly have suffered health problems. One major problem is mastitis, which is treated with antibiotics.[47] To maximize milk production, cows on tightly managed farms are impregnated as soon as two months after giving birth. That keeps them producing milk as though they were nursing, even though their calves usually are removed shortly after birth. Modern cows can sustain their extraordinary productivity for only about five or six years, at which time they are sold for beef. A well-cared-for cow normally could live into her 20s.[48]

The dairy industry is the source of at least 75 percent of the cattle that arrive at slaughterhouses unable to walk or stand.[49] Dairy farms produce so many "downers" because cows are slaughtered when their milk production falls, and decreased production usually occurs when the cows are either sick or old. Also, intensive milk production can deplete the calcium content of bones, increasing the risk that a cow will break a leg or pelvis. Downer

Growth Hormone: More Milk, Harm to Cows

Dairy farmers, ever eager to increase milk production, have turned for help to Monsanto's synthetic bovine growth hormone, Prosilac (also called recombinant bovine somatotropin, rBST). The hormone increases milk production by about 10 percent. However, it also increases the incidence of udder infections (mastitis) by about 25 percent, which may increase the need for antibiotics. A meta-analysis found that Prosilac also increases lameness by 50 percent, reduces fertility, and probably decreases cows' life spans.[50]

animals are frequently dragged into the slaughterhouse or lifted by a leg and hauled in.

Chickens

Today's hens produce an average of about 275 eggs per year, four times the 1933 average of 70. One of the key tools for maximizing production is the practice of forced molting. Under natural conditions, birds molt annually, shed-

ding and then replacing their feathers. During the process, egg laying slows to a halt, but the hen's reproductive tract regenerates, extending her productive life.

Natural annual molting is not efficient enough for farmers, so they induce molting by subjecting their chickens to stressors by restricting food (for up to 12 days), water (for up to 3), and sometimes light as well.[51] According to United Egg Producers' standards, birds subjected to such a regimen should lose no more than 30 percent of their weight and less than 1 percent should die.[52] The egg industry defends forced molting, stating that it increases hens' productive lives from 75 weeks to at least 110 weeks and decreases the number of new hens needed by 40 to 50 percent.

Development of Neurotic Behaviors

Cattle

Cattle on ranges walk several miles each day and spend 8 to 10 hours grazing. Confined cattle clearly cannot do this, and even lying down and standing up may be difficult in stalls. One response to this unnatural environment is that an animal will rub its head repeatedly against a stationary object, such as the bars of its cage, for extended periods. It might also bite the cage bars, grating its teeth back and forth on the metal. Cattle deprived of the ability to move freely sometimes roll their eyes back into their heads until only the whites are exposed.[53]

Additionally, the combination of implanted growth hormones, crowding, and the introduction of new cattle contributes to "buller syndrome" whereby one steer is ridden repeatedly by others in the group. That behavior is common in feedlots and may result in serious injuries, including broken legs or cracked spines.[54]

Other neurotic behaviors of cattle include stampeding, rejection of their young, and failure to produce milk.[55] While sometimes violent, the stereotyped behaviors of cattle are less often fatal than are those of pigs and chickens, as discussed below.[56]

Pigs

Pregnant sows confined in crates exhibit numerous abnormal behaviors.[57] They commonly chew while their mouths are empty, bite the bars of their cage, and constantly press the drinking nipple. Feed restrictions and boredom exacerbate those behaviors. Sows in gestation crates may sit on their haunches like 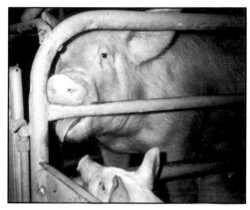 dogs—an atypical position for pigs, but one that they adopt because of the challenges of lying down and standing up in such limited spaces.

While pregnant, farrowing, and nursing sows are housed in tiny cages, most other pigs are kept in mid-sized group pens. The size of the pens and the number of animals in them varies from one operation to another, but when too many pigs are kept in a pen, they fare poorly. In a study of pigs subjected to a variety of stressors, being crowded with many other animals caused more stress than any other factor.[58] Pigs housed tightly together often bite each others' tails.[59] Once tail-biting begins, the behavior spreads rapidly through a herd. In some cases, the tail may be bitten down to the spinal cord, and some victims bleed to death or contract serious infections.[60]

Feral Pigs Versus Domesticated Pigs

The behaviors of domesticated pigs that are released into the wild contrast sharply with those of pigs raised in crowded indoor quarters. Feral pigs build small nests for group sleeping. They urinate and defecate at least 20 feet from their nests.[61] When wild sows become pregnant, they isolate themselves from the rest of their group and build a private nest in which to give birth. That behavior is impossible on factory farms, where pigs are trapped together and lack the materials for building nests. Also, feral pigs are highly social animals that typically live in family groups led by a dominant female.

Many of the abnormal behaviors of confined pigs—including tail biting—may be reduced simply by providing them with straw, sawdust, or other fibrous material.[62] Straw keeps floors drier and helps piglets stay warm. It also keeps animals from slipping, thereby reducing leg damage. Finally, it helps alleviate the tedium by allowing them to build nests and engage in other natural forms of behavior.

Chickens

Caged laying hens pace about to the extent they can and shake their heads in a neurotic manner. Those behaviors reflect the birds' perception of danger and their inability to escape. Under natural conditions, chickens instinctively search and investigate, pecking and scratching in the dirt for food most of the day. Denied anything to explore, caged hens exhibit polydipsia—the excessive manipulation of water dispensers and overconsumption of water.[63]

Chickens also resort to pecking their cage mates, leaving bare and bleeding patches on them and disrupting their ability to regulate their body temperature. In the worst cases, feather pecking results in death. While free-range chickens may peck one another, victims can escape, so injuries usually are less severe and fatalities less common.[64]

Confined layer hens obviously cannot engage in their natural nesting behavior because they do not have access to straw and other materials. Poultry will work hard to obtain nesting material and will go without food and water rather than without a nest.[65] Unable to perform that instinctive

Feral Chickens Versus Domesticated Chickens

Wild chickens serve as a useful indicator of how domesticated chickens might behave if industrial agriculture did not restrict their behaviors. Colorado State University professor Bernard Rollin has noted that feral hens forage over more than 100,000 square feet, while roosters cover five times as much ground.[66] That is in stark contrast to the 0.5 square feet available to caged layer hens. In the wild, the animals roost in groups of 6 to 30, with roosts positioned about 200 feet apart.

Feral hens demonstrate strong maternal behavior. For example, hens with chicks threaten other hens that come within 20 feet. Mothers do not start to leave their chicks until they are five to six weeks old. Farmed chicks, in contrast, are never mothered.

Farmed chickens still retain their ancestral instincts, judging from a study of chickens released into the wild. Amazingly, those highly inbred birds immediately began foraging for food, roosting in trees, building nests, and raising their young.[67]

behavior, chickens may "display agitated pacing and escape behaviors that last for two to four hours" before laying eggs.

For their part, broiler chickens, trapped in huge crowded houses, may exhibit "hysteria," a neurotic behavior marked by panicked vocalizing and wild flying.[68] That can lead to serious injuries.

Chickens also develop what is called deep pectoral myopathy.[69] The pectoral muscle normally is used to elevate wings, but in modern chickens it is rarely used. When the birds become excited—particularly when they are being chased and caught before transport—they suddenly and heavily exert that muscle. It expands within its thick, inelastic covering, cutting off its own blood flow. The muscle becomes dry and green and begins to die.

Some 60 million broiler chickens are raised each year strictly for the purpose of producing the next generation of broiler chickens.[70] Those hens' genetic makeup not only leads to fast growth, but also to heart disease and lameness. To avoid those problems and to increase fertility, producers underfeed their hens. Those birds are fed as little as half to a quarter of the amount of food they would otherwise eat. As a result, they are chronically malnourished and suffer psychological stress.[71]

Super Chickens

Broilers—chickens grown for meat—face unique challenges. Broilers once took 13 weeks to reach market weight, during which time they ate 3 pounds of feed for every pound of body weight gained. Losses primarily were due to infectious diseases, and mortality was as high as 30 percent. Now, modern breeding and feeding practices bring broilers to market weight in five to six weeks—and they need to eat only 1.8 pounds of feed to gain 1 pound of body weight. Mortality is only about 4 percent.[72] Those figures represent real progress, but the progress brings new problems.

Most deaths in broiler chickens now result not from infections or predators but from cannibalism in crowded chicken houses.[73] Also, skeletal growth cannot keep up with the extraordinary enlargement of muscle and body mass, so birds frequently suffer broken bones.[74] Chickens may die from obesity-related disorders, such as liver and kidney failure, or cardiovascular disorders.

What Farm Animals Consume

Animals raised on factory farms are used as living garbage disposals. In addition to grains and roughage, they may be fed newspaper, out-of-date baked goods, candy, industrial sludge, manure, and sewage, among other waste products.[75] Those "foods" may be contaminated with pesticides,

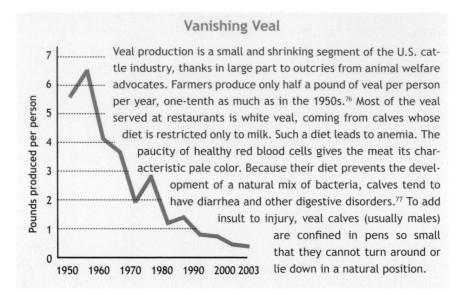

Vanishing Veal

Veal production is a small and shrinking segment of the U.S. cattle industry, thanks in large part to outcries from animal welfare advocates. Farmers produce only half a pound of veal per person per year, one-tenth as much as in the 1950s.[76] Most of the veal served at restaurants is white veal, coming from calves whose diet is restricted only to milk. Such a diet leads to anemia. The paucity of healthy red blood cells gives the meat its characteristic pale color. Because their diet prevents the development of a natural mix of bacteria, calves tend to have diarrhea and other digestive disorders.[77] To add insult to injury, veal calves (usually males) are confined in pens so small that they cannot turn around or lie down in a natural position.

heavy metals, and such carcinogens as polychlorinated biphenyls (PCBs), polybrominated biphenyls, dioxins, and furans, all of which are industrial by-products that pollute the environment.[78] In 2000, for instance, tests by the U.S. Food and Drug Administration (FDA) found that 44 percent of samples of animal feed contained pesticide residues, with 2 percent exceeding the legal limits.[79]

Some of the toxins livestock consume are fat soluble and build up in their body fat. The fattiest beef and dairy products (and, to a lesser extent, poultry and pork) deliver the highest concentrations of the toxins. Those chemicals also threaten the health of the animals, particularly just before and after birth, because during pregnancy and lactation a large portion of fat is mobilized in the mother's body. If the fat in the mother's milk contains toxins, newborns can experience significant exposures that may affect their health.[80]

You Call This Food?

Cattle, sheep, goats, and other ruminant animals evolved to eat and obtain energy from cellulose-rich grasses. That ability allows them to make use of plant matter that other animals cannot digest. However, cattle grow more slowly when they eat grasses than when they eat high-energy corn and other grains. So, to fatten their cattle as quickly as possible, ranchers typically ship them to feedlots for the final three to five months of their lives. There they are fed an unnatural diet that contains as much as 90 percent grain.

High-grain diets cause the gastrointestinal system to be more acidic. Normally, the rumen (the part of the stomach in a cow, sheep, or goat that digests grass and other food) is slightly acidic, with a pH near 6. On a high-grain diet, the pH may fall to 5 or even 4. A decrease of 1 pH unit means that the rumen is 10 times more acidic, and a decrease of 2 pH units means the rumen is 100 times more acidic.[81] The higher acidity alters the natural mix

of bacteria in the cattle's digestive system, selecting for bacteria that better tolerate acids. One such bacterium is *Escherichia coli* O157:H7, the nasty foodborne pathogen that causes about 80 deaths annually in the United States (see appendix A, p. 172).[82]

The digestive system of cattle is not designed to process large amounts of grain. The result: ulcers, bloat, liver abscesses, hoof infections, growth of acid-tolerant E. coli O157:H7.

The altered bacterial environment can cause ulcers in the rumen. Bacteria then may travel through the ulcers to the liver, where they frequently cause abscesses. To help prevent that, feedlot operators add antibiotics such as tylosin (an antibiotic similar to the erythromycin used to treat infections in humans) to the animals' feed. Without antibiotics, the livers from about 75 percent of cattle would have to be discarded due to abscesses. Even with antibiotics, about 13 percent of livers are condemned at slaughter.[83]

Bacteria that migrate through ulcers also can infect the hooves of cattle and cause lameness, which accounts for 16 percent of feedlot health problems and 5 percent of deaths.[84] Less commonly, a high-grain diet causes dehydration, shock, and kidney failure.[85]

An acidified rumen can trigger diarrhea, bloat (likened by one expert to "a massive stomach ache"), and grain overload, a potentially fatal condition.[86] James Russell, a Cornell University and USDA microbiologist, estimates that about 3 of every 1,000 cattle in feedlots die of grain-related disorders.[87]

Grain-based diets must be introduced gradually to cattle. When cattle are fed too much grain too suddenly, their rumens may develop bloat, expanding to the point where they press up against the lungs. Without immediate medical attention, bloat can cause death by suffocation.[88] That is what happened in 2005 at a feedlot in Alberta, Canada. Because the cattle's feed was incorrectly mixed and contained too much barley and barley silage, it caused "acute carbohydrate ingestion" and killed 150 cattle.[89]

High-grain diets appear to cause abnormal behaviors. In pursuit of roughage, cattle will chew on any available source, including wooden fences.[90] The lack of roughage, coupled with confinement, also results in neurotic tongue rolling. That behavior simulates the motion of wrapping the tongue around tufts of grass—except that the grass is imaginary.[91] Grazing cattle curl their tongues around tufts of grass thousands of times per day, so it is not surprising that the absence of such a customary behavior has consequences. Tongue rolling is not observed among cattle on pasture or among wild bovines.

Antibiotics, Antacids, and More

Industrial agriculture—with its dirty, overcrowded, high-production systems—often increases the likelihood of certain illnesses and the need for antibiotics. As mentioned earlier, intensive milk production causes mastitis in the udders of dairy cows, which is treated with antibiotics. Hogs in cramped pens may bite off one another's tails, leaving exposed sores ripe for infection. Crowding also speeds the spread of disease among animals. And processing animal wastes (discussed below) into feed transmits pathogens to animals and increases the risk of infections and need for medication.[92]

Feedlot operators use antibiotics and antacids to prevent and treat the diseases caused by high-grain diets. The antibiotics are added to cattle feed to kill the bacteria that cause liver abscesses and hoof infections (see "Factory Farming's Antibiotic Crutch," p. 68). Feedlot operators also add ordinary baking soda, limestone, and other alkaline substances to feed to neutralize excessive acidity in the rumen.[93] In fact, one-fourth of all baking soda produced in the United States is fed to livestock.[94]

Pesticides and Other Chemical Toxins

PCBs and organochlorine pesticides are "endocrine disruptors," which may strengthen or weaken the action of natural hormones in animals and humans. Endocrine-disrupting compounds (EDCs) may affect many aspects of development, but especially sexual development. For example, they decrease sperm production in many animals, including humans. In cattle, EDCs can upset the maturation of oocytes (the cells that produce eggs in female mammals). During sensitive periods of development, EDCs can induce physiological changes at concentrations less than one-hundredth of those that are toxic at other times.[95]

Sex Hormones on Ranches

Ranchers routinely castrate their bulls to make them more docile and produce fattier, more tender meat, but castration reduces the levels of growth-promoting hormones and the growth rate of steers. For 25 years the cattle industry has compensated for that slower growth by implanting natural (estradiol, progesterone, testosterone) or synthetic (trenbolone acetate, zeranol) sex hormones into steers' ears (which are discarded after slaughter). Another hormone, melengestrol acetate, is added to feed. A dollar's worth of hormones saves at least 10 dollars' worth of feed.[96] Pigs and chickens are not allowed to be treated with hormones.

The use of hormones is controversial, because the slightly increased amounts of hormones in beef could conceivably affect growth and development or cause cancer in consumers. In 1989 the European Union banned hormone-treated beef, including imports from North America. (The ban hasn't stopped many European farmers from injecting hormones illegally.) European officials contend that hormones—especially estradiol—might cause cancer or neurological, developmental, reproductive, or immunological effects.[97]

In fact, there's little evidence that hormones pose a risk.[98] The World Health Organization explains that hormone implants in treated cattle "contributed only a small additional amount of hormone to the intakes resulting from consumption of other foods." Indeed, we ingest far more hormones from eggs, milk, and soybean oil than from meat. The FDA and USDA note that hormone levels in meat from treated cattle are within the normal range of untreated animals. Moreover, very little of the hormones in beef is absorbed by the body, and, in any case, even children produce far more hormones than are present in meat. Finally, one marketer of

both treated and untreated beef acknowledges that the hormone-free claim is a "marketing tool used to create a false fear."[99] While one can't prove that anything is perfectly safe, hormone implants (especially the natural ones) do not appear to be worrisome.

Separate from health concerns, toxicologists have discovered that hormones in the manure of feedlot cattle (and urban sewage treatment plants) can pollute nearby streams.[100] The hormones, both naturally occurring and the extra amount from implants, are associated with smaller testes and fewer offspring in minnows and might also affect other wildlife. Edward Orlando, a reproductive physiologist at Florida Atlantic University, worries that "we know almost nothing about the environmental impact of hormones from agricultural sources."[101] The solution would be to reduce the concentration of cattle at feedlots and prevent water pollution from manure and urine.

Dioxin causes cancer in animals, as well as reproductive and developmental problems. In 2003, dioxin-contaminated waste from a brass factory was inadvertently used in animal feed, causing numerous deaths.[102] The contamination eventually was discovered, and the remaining meat and milk from the animals that had eaten the feed were removed from the market. However, a good deal of dioxin must have passed from meat and dairy products to consumers before the problem was identified.

Sludge, Manure, and Feces

Sewage sludge—everything pulled out of dirty water at waste-treatment plants—and raw or composted manure are commonly fed to livestock both directly and indirectly. For example, manure is commonly applied to cropland because it contains nutrients that serve as fertilizer. But manure is also rich in bacteria and chemical contaminants, including organic compounds and heavy metals. Livestock are exposed to those hazards when they graze on manure-treated fields and when they eat crops grown in contaminated soil. Since 1992, when Congress banned ocean dumping of sewage sludge, municipalities' most common method for disposing of sewage waste has been to offer it to farmers as fertilizer.[103] Excess nitrogen from the waste can pollute the water drunk by calves and lambs, causing "blue baby syndrome," which on rare occasions results in suffocation and death.[104] That syndrome, which also occurs (albeit rarely) in humans, develops when young animals consume excess nitrate, which, when converted by bacteria in the gut to nitrite, prevents hemoglobin from carrying oxygen.

Candid Camera at a Poultry Farm

In 2004, workers at a West Virginia facility owned by Pilgrim's Pride—the second-largest poultry producer in the United States—were caught on videotape stomping on live chickens, throwing them against walls, and kicking them. Although not typical, that despicable behavior serves as a useful reminder that poultry are not protected by the Humane Methods of Slaughter Act for livestock, leaving them open to a variety of cruel practices.[105]

"Poultry litter"—a euphemism for the mixture of manure, feathers, wood chips, and spilled feed collected from the floors of poultry houses—is commonly fed to other animals. In one extreme case, poultry litter contaminated with high levels of copper, which is used to control everything from algae to snails, killed cattle and sheep.[106]

Animal Parts

Feeding animal parts back to animals exacerbates the risks from all of the contaminants to which they are exposed. Although a federal law forbids feeding cattle parts back to cattle, rendered farm animals remain a common component of livestock feed. Rendered poultry and hogs may be fed to cattle, and rendered cattle to poultry and hogs.[107] Thus, livestock consume the same concentrated toxins from the fat of slaughtered animals as meat and dairy eaters. That process may increase contamination of animal products over time and keep banned or unused chemicals circulating in the food supply. Moreover, that cycle of feeding animals to animals may be a route for transferring mad cow disease (for more on that topic, see appendix A, p. 174).

How They're Transported

While some animals may take only one trip during their lives—to the abattoir—most cattle and pigs endure the stress of transport several times. According to the USDA, only 24 percent of sheep and 29 percent of pigs grow up on the farm where they are born.[108] In a single year, 22 million cattle and 27 million hogs were shipped to another state to be fattened or bred.

The trucks and railroad cars used to transport livestock from farm to farm or to feedlots and slaughterhouses are even more cramped than the factory farms themselves. On a truck, the space recommended by animal welfare experts for a 1,000-pound cow is only 12.8 square feet—half of what typically is provided in a feedlot.[109] Using the Mini Cooper analogy again, six 1,000-pound cows are packed into an area that that petite car occupies. Full-grown, 1,400-pound cattle get only 19 square feet. A 400-pound hog is allotted 6½ square feet in a truck. That's half the size of a gestation crate; almost 12 hogs could fit into a Mini's footprint.

Chickens are "harvested" roughly by "catchers" who cram them into crates, which are then trucked to the slaughterhouse.

Eighty percent of the calves from Texas and Kansas are shipped an average of about 200 miles before they are killed.[110] During transit, animals are generally deprived of food and water. Under those conditions, chickens sometimes suffer heart failure, pigs die from the cold, and sheep may be smothered.[111] And the extreme temperatures—just think of traveling jam-packed in a slat-sided tractor trailer on a Texas highway in the August heat—kill some animals outright.[112]

Shipping promotes the spread of disease among animals, especially when animals from different herds and flocks exchange pathogens. That is compounded by the stresses of extreme temperatures, an unfamiliar environment, and forced crowding, which can suppress the animals' immune systems.[113]

Movement is particularly difficult for cattle because, according to Colorado State University's Bernard Rollin, they "are creatures of habit, and disruption of habits can be highly stressful.... Introduction into a new environment is more stressful for cattle than electric shock."[114] In cattle, the most common result of stress and exposure to germs is "shipping fever,"[115] also called bovine respiratory disease complex. That is a severe form of pneumonia and the most common cause of death in factory-farmed cattle. Severe outbreaks, though rare, have killed up to 35 percent of a herd.[116]

Rough handling injures broiler chickens. When catching chickens for their journey to the slaughterhouse, workers typically carry up to seven at a time by one leg. That frequently results in broken bones and dislocated wing and leg joints.[117]

How They're Slaughtered

After the miserable lives most farm animals lead before reaching the slaughterhouse, one would hope that their deaths would at least be quick and painless. Unfortunately, that sometimes is not the case. Animals may endure inhumane conditions while waiting at the slaughterhouse, and,

Candid Camera at a Cattle Slaughterhouse[118]

Ritual slaughter of animals to provide kosher meat involves cutting the animals' necks without stunning them first. A kosher slaughterhouse operated in Pottsville, Iowa, by AgriProcessors, Inc., was caught on videotape apparently violating both kosher laws and the Humane Slaughter Act. The aggressive animal rights group, People for the Ethical Treatment of Animals (PETA), secretly videotaped mistreatment. As the *New York Times* described tape, "after steers were cut by a ritual slaughterer, other workers pulled out the animals' tracheas with a hook to speed bleeding. In the tape, animals were shown staggering around the killing pen with their windpipes dangling out, slamming their heads against walls and soundlessly trying to bellow. One animal took three minutes to stop moving."

Abattoirs: Hell for Workers, Too

In July 2000, Jesus Soto Carbajal, a worker at a Cargill meatpacking plant in Nebraska, was cutting hindquarters of beef "coming down the line at him every six seconds."[119] Eventually, the fast pace caught up with Carbajal when his knife slipped and sliced open his jugular vein. He died almost immediately.

Contemporary meat and poultry slaughterhouses and processing plants are America's most dangerous places to work. The average slaughterhouse worker is three times more likely to be injured than the average factory worker.[120] The handling of large and frightened animals, the use of dangerous equipment, and the inadequate training of workers all contribute to the epidemic of injuries.

The demand for speed at slaughterhouses and processing plants creates a dilemma for workers: Accept unsafe working conditions or risk being fired.[121] Speed leads to carelessness, increasing workers' risk of injury or death. For instance, when workers fail to completely stun cattle or hogs, the animals can regain consciousness and attack the workers.[122]

The greatest workplace dangers include high-speed processing lines, sharp knives, heavy lifting or pushing of animal carcasses, dangerous bacteria from animal remains, long hours and mandatory overtime, poor training, and a lack of protective clothing and ergonomically safe equipment.[123] Additionally, a lack of union representation removes a strong force for improving working conditions.

Human Rights Watch characterizes meatpacking and slaughtering plants as places "where exhausted employees slice into carcasses at a frenzied pace…often suffering injuries from a slip of the knife or from repeating a single motion more than 10,000 times a day."[124] In *Fast Food Nation*, Eric Schlosser describes crippling injuries or death, including ones documented by the Occupational Safety and Health Administration: plant workers having hands or other limbs crushed or severed by machinery, workers being pulled by conveyor belts into grinding equipment, slippery floors causing workers to fall from great heights to their deaths, and workers suffering asphyxiation from cleaning blood collection tanks filled with toxic gases.[125]

despite companies' intentions, their deaths may be slow and excruciating. Those conditions also endanger slaughterhouse workers (see "Abattoirs: Hell for Workers, Too," above). The World Organization for Animal Health, which is supported by 167 member countries, offers detailed recommendations for transporting and slaughtering animals in a humane way, but governments must implement that advice.

Some holding areas at slaughterhouses have no water. That means animals are unable to drink from the time they are first loaded onto the trucks until the time they are slaughtered. Many plants use electric prods to drive cattle into the slaughterhouse.[126]

The rate of killing in a typical modern slaughterhouse is breathtaking: 13,200 chickens per hour, 1,100 pigs per hour, 250 cattle per hour. At those speeds, it is likely impossible to ensure that all animals have been adequately stunned before they are killed. According to a report commissioned by the USDA, in some plants, as many as 8 percent of pigs, 20 percent of cattle, and 47 percent of sheep were not properly stunned.[127]

Slaughterhouses stun broiler chickens by placing their heads in an electrified pool of water. After that, the chickens' necks are slit. Layer hens are not typically stunned, because their osteoporotic, unexercised bones break when exposed to the electrical current.[128] Instead, layers are conscious while their throats are slit.

Cattle and pigs usually are stunned by a pneumatic bolt shot into their foreheads. But some cattle are not stunned properly and "are often still alive and conscious as they proceed down the production line," according to the Humane Farming Association, an animal welfare organization.[129] In both kosher and halal slaughter, cattle or chickens typically are not stunned before their throats are slit, so they are fully conscious when they are cut and as they bleed.[130]

"The chicken industry is way behind the beef and pork industries" in terms of adopting more humane practices, according to Professor Temple Grandin of Colorado State University, a widely respected expert on animal slaughter techniques.[131] Although figures are not available for the United States, in Europe—where similar slaughter methods are used—about 30 percent of broiler chickens are not adequately stunned before slaughter.[132] That means the animals may suffer extreme pain as they are being "deconstructed." Poultry are exempt from the Humane Methods of Slaughter Act of 1978, which requires livestock handling and slaughtering to be "carried out only by humane methods" and calls for animals to be "quickly rendered insensible to pain before they are slaughtered." The Humane Society of the United States has sued the USDA to end that exemption.[133]

Agriculture Also Affects Non-Farm Animals

Livestock are not the only animals that suffer from the current system of agricultural production in the United States. The USDA acknowledges that agricultural practices are the "primary factor depressing wildlife populations in North America."[134] Of the 663 species listed as threatened or endangered, 272 made the list because of agricultural expansion and 115 due to the use of fertilizers and pesticides.[135] Agricultural pesticides are associ-

Sport fish, such as this trout, can be harmed by pesticides. They can also concentrate toxins in their fat and harm their predators—both human and animal.

ated with possible changes in hormonal activity in frogs and other amphibians (see "Pesticides: Gauging the Health Risk," p. 84). Agricultural fertilizers and livestock manure pollute streams, causing algal blooms that starve fish of oxygen, ultimately suffocating them (see "Modern Farming Practices Pollute Water," p. 94). Sewage sludge applied to pasture not only affects livestock, but deer and other animals as well. As with livestock, the toxins in the sludge may kill the animals outright or may be stored in their fat and passed on to the dwindling number of large predators, such as wolves. A more direct threat is that the U.S. Fish and Wildlife Service and state governments routinely kill wolves that may threaten livestock.

Pesticides—especially insecticides—unintentionally kill many species, including the natural predators of crop pests. (It is important to note that what might be considered a pest to crops could be beneficial in a different ecosystem. Thus, the term "pest" does not necessarily mean an organism is inherently harmful.) Each year, millions of pounds of pesticides lethal to a broad range of species are applied across millions of acres of farmland. That usage causes widespread ecological harms to non-target species, such as insects, weeds, fish in nearby rivers and streams, and the wildlife that depend on those insects and fish for survival.

Consumers higher up on the food chain—including insects, birds, larva-eating fish, frogs and other amphibians, and other terrestrial mammals—

may be poisoned by consuming pesticide-contaminated prey. Because those consumers typically reproduce in smaller numbers than the insects and other creatures they eat, their populations are less capable of recovering. Once those populations are reduced, more pesticides are required to control the pests that the predators otherwise would have controlled.

Some species are particularly vulnerable to pesticides. For example, the relatively large surface area of small insects allows them to absorb lethal doses quickly and easily, making non-target insects the frequent victims of pesticide poisonings.[136] Honeybees, for example, have been so devastated by factors including pesticides that many farmers need to rent beehives to ensure that their crops are pollinated. The decline in important pol-  linators has led the American Beekeeping Federation to decry the overuse of pesticides, and the North American Pollinator Protection Campaign is actively trying to reduce the misuse of pesticides that kill insects that pollinate crops.[137]

Birds, fish, and other wildlife are exposed to agricultural-use pesticides that remain in the environment. Direct or indirect contact with those pesticides can poison them.[138] Pesticides on U.S. farmland have been estimated to kill about 67 million birds each year.[139] With birds, for example, the more they feed on fish that have ingested pesticides (as a result of runoff from contaminated soil or drift), the more pesticides those birds accumulate in their tissue.

Fish, too, are highly susceptible to pesticides as a result of soil runoff from farmland into bodies of water or from drift during or after pesticide applications. Major fish kills have been attributed to aerial sprayings of herbicides and insecticides on farmland. High levels of those pesticides were found in the fish that survived.[140] Because fish are important prey for many species of birds, contaminated fish harm birds and threaten ecosystems.

What It All Means

Humane treatment of livestock should be an ethical imperative. Giving animals enough space so that they are not driven to attack each other is not difficult—farmers provided that for generations. Allowing animals to act out most of their natural behaviors should be achievable. If they were allowed to go outside, given straw so they could build nests, and permitted to establish a natural social order, fewer animals would be needlessly injured or killed. Avoiding certain cruel procedures altogether makes sense—especially when they are only performed because of inappropriate animal husbandry (as with debeaking chickens and detoeing turkeys). Feeding cattle diets that do not make them sick is feasible—let them eat what they always ate, instead of fattening them on grain and toxin-tainted feed. When animals are shipped, they should be given adequate room and protection from extreme heat or cold—that is done for horses all the time. Finally, animals could be slaughtered humanely if workers were adequately trained, slaughtering lines were slowed down, and poultry were rendered unconscious by inert gases.

Until practice is consistent with theory, the simplest thing a consumer could do for animal welfare is to eat less (or even no) meat and other animal products. That would reduce the number of farm animals and the potential for mistreatment. Consumers also could choose meat and dairy products made from more humanely raised animals (see www.certifiedhumane.org or www.eatwild.com). Meanwhile, the entire animal-food industry—voluntarily or in response to new laws—should be improving its practices as much as possible. While those improvements might raise the price of animal products, the higher prices we would pay at the grocery store would be slight indeed compared to the price livestock are now paying.

Making Change

Changing Your Own Diet

As you've now seen, what you eat has effects that ripple not just through every organ in your body, but also through the natural environment and farms and other parts of the food industry. Will you change your diet or continue eating as you have been? If you're like most of us, you probably could make some easy and tasty changes that will help protect your arteries, protect the planet, and protect farm animals.

While some people think nutrition is impossibly complicated, today's basic dietary message is actually quite clear and simple. The experts (see table 1) recommend that you:

- Base your diet largely on vegetables, fruits, beans, whole grains, and healthy oils.
- Eat fish and only modest amounts—if you choose to eat them—of fat-free or low-fat meat and dairy products.

Table 1. **Health experts' dietary advice**[1]

Organization	Nutrition advice
American Cancer Society	"Eat five or more servings of a variety of vegetables and fruits each day....Limit consumption of red meats, especially those high in fat and processed [bacon, ham, sausage]. Choose fish, poultry, or beans as an alternative to beef, pork, and lamb."
American Diabetes Association	"Reduced intake of total fat, particularly saturated fat, may reduce risk for diabetes...[as would] increased intake of whole grains and dietary fiber."
American Heart Association	"Consume a diet rich in vegetables and fruits...whole-grain, high-fiber foods...fish...lean meats and vegetable alternatives, fat-free (skim) or low-fat (1% fat) dairy products."
American Institute for Cancer Research/ World Cancer Research Foundation	"Choose predominantly plant-based diets, rich in a variety of fruits and vegetables, pulses (legumes), and minimally processed starchy foods."
2005 Dietary Guidelines for Americans	"A healthy eating plan is one that emphasizes fruits, vegetables, whole grains, and fat-free or low-fat milk and milk products. Includes lean meats, poultry, fish, beans, eggs, and nuts. Is low in saturated fats, trans fat, cholesterol, salt (sodium), and added sugars."
World Health Organization	"[Eat] more fruit and vegetables, as well as nuts and whole grains....[Cut] the amount of fatty, sugary foods in the diet....[Move] from saturated animal-based fats to unsaturated vegetable-oil based fats."

• Cut way back on salt, refined sugars, white flour, and partially hydrogenated oils.

Making the right dietary choice can be extended beyond health concerns by eating in an environmentally responsible way. Raising livestock requires far more resources—land, energy, pesticides, fertilizer, and water—and generates far more pollution than growing fruits, vegetables, and grains. Among animal products, producing grain-fed beef harms the environment much more than raising poultry and grass-fed beef and producing dairy foods.

In addition, we should consider animal welfare. Out of sight is usually out of mind, and it is all too easy to forget about cramped chicken coops, filthy slaughterhouses, and the like when we sink our teeth into a juicy charbroiled steak or grilled chicken breast. Those considerations suggest the benefits of not only avoiding *fatty* meat, dairy foods, and poultry, but of eating *less* animal products and getting essential nutrients from other sources. Far from being a punishment, eating such a diet opens up a multitude of wonderful new taste sensations. Alternatively, you could make a special effort to buy meat, dairy products, and eggs from humanely raised animals, ideally from small, local farms.

We can all take control of our diets, even in a culture that encourages people to eat hamburgers, hot dogs, and soda pop almost from birth. Two healthy diets that are easy to follow—and delicious—are the modified Dietary Approaches to Stop Hypertension (DASH) Eating Plan (see figure 1) and the Mediterranean Food Pyramid (see figure 2). The DASH Eating Plan was developed by the National Institutes of Health for studies on blood pressure.[2] It is loaded with fruits and vegetables; recommends nuts, seeds, and low-fat dairy foods; and includes modest amounts of fish and low-fat meat and poultry. It also is low in sodium. The DASH diet includes no more than 4 to 5 servings of low-fat animal foods, and 16 to 19 servings of plant foods, per day.

The Mediterranean Food Pyramid (figure 2) was developed by Oldways, a nonprofit organization that advocates healthy, traditional diets. It is based

Figure 1. **The DASH Food Pyramid**[3]

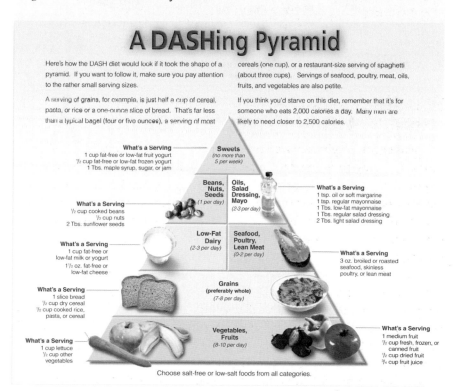

Choose salt-free or low-salt foods from all categories.

Figure 2. **Healthy Mediterranean Diet Pyramid**

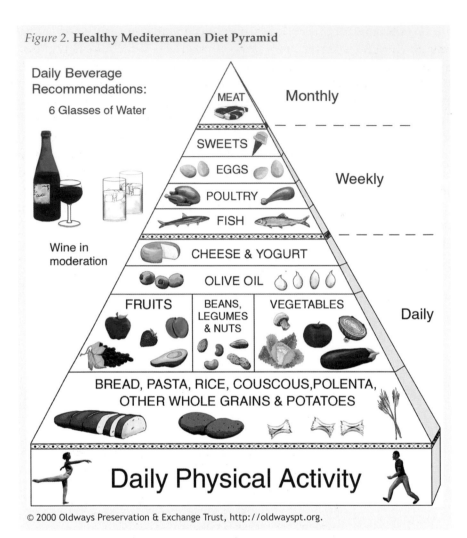

on the diet once consumed widely in southern Europe, including mainland Greece, the island of Crete, and southern Italy. The diet includes modest amounts of dairy foods, fish, poultry, and eggs; wine in moderation; and plenty of fruits, vegetables, beans, and whole grains. The Mediterranean diet allows red meat only rarely. Both the DASH and Mediterranean diets specify much less refined sugars than most Americans eat, and they pretty much exclude butter and stick margarine. Oldways' Healthy Mediterranean Diet Pyramid also emphasizes daily physical activity, but that's a given with any diet. Walking, biking, jogging, tennis, swimming, weight-lifting, and other activities are essential to good health.

For those who have ethical concerns about animal welfare or eating animal products, a healthy vegetarian or vegan diet is the way to go. Such a diet is based on fruits, vegetables, whole grains, dried beans, nuts, and, if not vegan, low-fat and non-fat dairy products and egg whites. Either a lacto-ovo vegetarian diet or a vegan diet can provide all the necessary nutrients while minimizing the risk of chronic disease.[4] For any doubters, the American Dietetic Association and Dietitians of Canada are reassuring: "Appropriately planned vegetarian diets are healthful, nutritionally adequate, and provide health benefits in the prevention and treatment of certain diseases."[5] Those groups produced a Vegetarian Food Pyramid that features a healthy lacto-ovo vegetarian diet (see figure 3 for our adaptation of that pyramid).[6]

Figure 3. **Vegetarian Food Pyramid**[7]

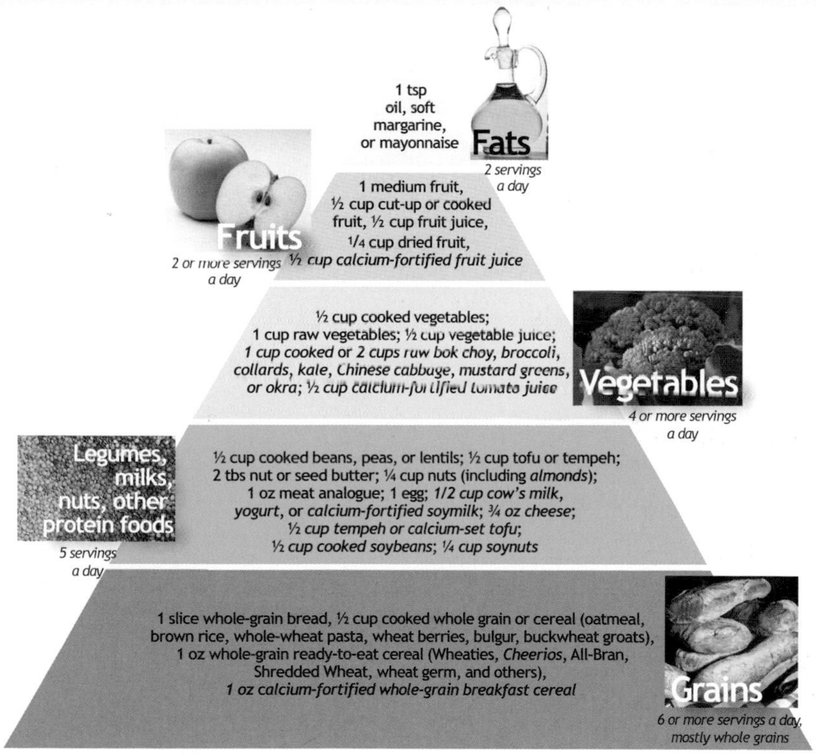

1 tsp oil, soft margarine, or mayonnaise **Fats**
2 servings a day

1 medium fruit, ½ cup cut-up or cooked fruit, ½ cup fruit juice, ¼ cup dried fruit, *½ cup calcium-fortified fruit juice*
Fruits
2 or more servings a day

½ cup cooked vegetables; 1 cup raw vegetables; ½ cup vegetable juice; 1 cup cooked or 2 cups raw bok choy, broccoli, collards, kale, Chinese cabbage, mustard greens, or okra; ½ cup *calcium-fortified tomato juice*
Vegetables
4 or more servings a day

½ cup cooked beans, peas, or lentils; ½ cup tofu or tempeh; 2 tbs nut or seed butter; ¼ cup nuts (including *almonds*); 1 oz meat analogue; 1 egg; *1/2 cup cow's milk, yogurt, or calcium-fortified soymilk; ¾ oz cheese;* ½ cup tempeh or calcium-set tofu; ½ cup cooked soybeans; ¼ cup soynuts
Legumes, milks, nuts, other protein foods
5 servings a day

1 slice whole-grain bread, ½ cup cooked whole grain or cereal (oatmeal, brown rice, whole-wheat pasta, wheat berries, bulgur, buckwheat groats), 1 oz whole-grain ready-to-eat cereal (Wheaties, *Cheerios*, All-Bran, Shredded Wheat, wheat germ, and others), *1 oz calcium-fortified whole-grain breakfast cereal*
Grains
6 or more servings a day, mostly whole grains

Notes: This pyramid is designed for both vegans and those who eat dairy products and eggs. Aim for eight servings a day of calcium-rich foods (in *italics*), and be sure to get sufficient vitamins B12 and D from foods or supplements. (A serving of milk or yogurt is ½ cup.)

The Vegetarian Food Pyramid replaces meat and poultry with nuts (and nut butters), beans (including tofu), seeds, and eggs. It emphasizes low-fat or non-fat milk, yogurt, or cheese or vegetarian substitutes. Vegans can easily adapt that lacto-ovo diet to their needs. Because of their more restricted diets, vegetarians (especially vegans) should eat fortified foods or take dietary supplements to ensure that they consume adequate amounts of vitamin B12, calcium, vitamin D, iron, and zinc.[8]

> ## Avoid Food Poisoning
>
> Whatever diet you choose, protect yourself from germs that lurk in animal products *and* in fruits and vegetables. Wash your hands and all cooking implements after they come in contact with raw meat and poultry. Wash fruits and vegetables before eating them. And keep hot foods hot and cold foods cold.

Anyone who does continue to eat animal foods should consider buying ones that caused the least misery for the animals. That means eggs from uncaged hens, beef from cattle that never saw a feedlot, pork and poultry from pigs and birds that could roam about, and milk from cows that grazed on pastures, weather permitting. Look for label claims like "humanely raised" and, if you're at a farmer's market, ask the farmers about their practices. Even animals raised organically are not necessarily raised in the most humane ways. Several resources are provided in appendix B, p. 179.

When changing your diet for health, environmental, or ethical reasons, you need to remember that avoiding fatty meat and dairy products is only half the solution. The other half is choosing *healthy* plant-based foods. Most of the bread, pasta, rice, and other grain foods that Americans consume are made from refined grains; soft drinks and candy are made with empty-calorie sugar and high-fructose corn syrup; and too much once-healthy vegetable oil has been partially hydrogenated and contains artery-clogging trans fat (that is especially the case for fried foods at restaurants).

Making several little changes quickly adds up to an overall healthier diet. Consider someone who replaced one 3½-ounce serving of beef, one egg, and a 1-ounce serving of cheese each day with a mix of vegetables, fruit, beans, and whole grains. That modest change would increase the person's daily consumption of dietary fiber by 16 grams (more than half the recommended intake) and reduce the intake of fat by 22 grams (one-third of the recommended daily limit) and saturated fat by 12 grams (more than half the recommended limit).[9] In environmental terms, over a year, those changes would spare the need for 1.8 acres of cropland, 40 pounds of fertilizer, and 3 ounces of pesticides. It also would mean dumping 11,400 fewer pounds of animal manure into the environment. Multiply those improvements by millions of people and it's easy to see the dramatic improvements in health and reductions in pollution that dietary changes could bring about. (You can see the effects of the dietary changes that *you* might make by using our computerized calculator at the Center for Science in the Public Interest's web site at www.EatingGreen.org.)

We've created a Diet Scorecard (next page) to help you gauge the overall impact of your diet on your health, the environment, and the welfare of farm animals (our computerized version is a lot easier to use). The health score reflects the benefits of plant foods, seafood, and low-fat animal products and the harm from the saturated fat, cholesterol, and other substances in animal foods. The environmental dimension considers such factors as air and water pollution from feedlots and industrial-style hog and poultry production, methane emitted by cattle, and problems related to fertilizers and pesticides. The animal welfare score reflects such practices as crowding on factory farms and feedlots and inhumane treatment at slaughterhouses.

The Diet Scorecard

Pencil in the number of servings you eat each week, then calculate your health, environmental, and animal-welfare scores. Total each column, then tally your grand total. Carefully note the + and - signs, which tell you whether to add or subtract points. A computerized version of Diet Scorecard is available at www.EatingGreen.org.

Food (average diet)	Servings YOU eat per week	Health score	Environmental score	Animal welfare score
Beef (4 servings/week)		− (2.5 × servings)	− (2 × servings)	− (1 × servings)
If you eat mostly grass-fed beef, add points		+ (servings ÷ 2)	+ (servings ÷ 2)	+ (servings ÷ 2)
Pork (3 servings/week)		− (2 × servings)	− (2 × servings)	− (2 × servings)
If you eat mostly extra-lean beef or pork, add points		+ (1 × beef + pork servings)		
Chicken, turkey (non-fried) (4 servings/week)		+ (servings ÷ 2)	− (servings ÷ 2)	− (2 × servings)
If you eat mostly free-range poultry, add points			+ (poultry servings ÷ 4)	+ (1 × poultry servings)
Egg yolks (5 yolks/week)		− (servings ÷ 2)	− (servings ÷ 2)	− (2 × servings)
If you eat mostly free-range eggs, add points			+ (above points ÷ 4)	+ (3/4 × above points)
Milk, yogurt (7 cups/week)		− (1.5 × servings)	− (1 × servings)	− (servings ÷ 2)
If you mostly or sometimes eat fat-free/low-fat dairy products, add points		+ (mostly: add 2 × above points) (sometimes: add 1 × above points)		
Cheese (Swiss, American, other hard cheeses) (4 oz. or 4 slices/week)		− (3 × servings)	− (2 × servings)	− (1 × servings)
If you normally eat low-fat cheese, add points		+ (1.5 × cheese servings)		
Fish (1 serving/week)		+ (4 × servings)	− (servings ÷ 2)	− (servings ÷ 3)
Fruit (7 servings/week)		+ (3 × servings)	− (servings ÷ 4)	
Vegetables, beans, nuts (omit potatoes, iceberg lettuce) (15 servings/week)		+ (3 × servings)	− (servings ÷ 4)	
Whole grains (7 slices bread/week)		+ (1.5 × servings)	− (servings ÷ 4)	
Candy, pastries, hotdogs, ice cream, etc.		− (1 × servings)		
If you eat mostly organic food, add points		+ (mostly: add 2 points) (sometimes: add 1 point)	+ (mostly: add 2 points) (sometimes: add 1 point)	
Your totals:				
Grand total (add 3 totals above):				

60+: Excellent!; 15-59: Good; Below 15: Uh-oh, you need help!

Changing Government Policies

The preceding chapters have detailed many of the human health, environmental, and animal welfare problems stemming from animal agriculture—particularly when conducted on an industrial scale. All of those problems would be diminished if Americans switched to a more plant-based diet.

Although millions of people have adopted healthier, more plant-based diets, change is hard, because diet is embedded in our family traditions and culture and perpetuated by major industries. It will take more than occasional public service messages, newspaper articles, and official reports to get the bulk of the population eating a "greener" diet.

This chapter suggests a variety of government programs and policies—few of which would be easily obtained—that would help move Americans toward a more plant-based diet. Recognizing that not everyone would or should become a vegetarian, we suggest means of both obtaining healthier animal products and improving how animals are raised. Consumer demand will be the most important factor in changing what people eat, what food marketers offer, and what farmers grow. But nutrition- and environment-based food and farm policies could improve diets indirectly. To that end, some of the policy options suggested here would "internalize" the health and environmental costs of producing animal products. That would mean paying a little more at the supermarket, but paying less in the form of higher medical costs and a degraded environment. As Joel Salatin, a Virginia farmer who is a passionate advocate of small farms and local agriculture, is quoted in Michael Pollan's *The Omnivore's Dilemma*, defending the sometimes higher prices small farmers charge:

> I explain that with our food all of the costs are figured into the price. Society is not bearing the cost of water pollution, of antibiotic resistance, of foodborne illnesses, of crop subsidies, of subsidized oil and water—all of the hidden costs to the environment and the taxpayer that make cheap food *seem* cheap.[1]

Our focus here is on government actions, but companies could act a lot faster voluntarily. Some progressive companies and farmers, large and small, alternative and mainstream, already are producing healthier foods, minimizing their impact on the environment, and raising animals humanely. We hope other companies will emulate them.

One activity not discussed below, but important, is research. It is crucial that government continues to invest generously in objective scientific and economic research on health, the environment, and animal welfare. That research will provide insights on the effects of different diets and farming methods and suggest ways to improve government policies and industry practices.

Improving Human Health

The federal government invests billions of dollars a year in the food stamp program, school lunches and breakfasts, and similar programs. It feeds millions daily at cafeterias in its hospitals, mess halls, office buildings, and prisons. It spends tens of millions of dollars a year on nutrition research and provides sensible nutrition advice. But the government makes poor use of its knowledge, resources, and facilities when it comes to preventing heart disease, diabetes, and other diet-related diseases and saving the tens of billions of dollars that are now wasted on treating those often-preventable diseases. The recommendations outlined here challenge the government to put its words into action.

1. Increase Fruit and Vegetable Consumption

Consuming more nutrient-dense fruits and vegetables is one of the most important dietary changes that consumers should make. Eating more fruits and vegetables is heartily endorsed by the *2005 Dietary Guidelines for Americans*, because doing so would add vital nutrients to diets and could displace less-healthful foods. The government should show that it means what it says by sponsoring programs, including the following, that would have a real impact.

- Intensive media campaigns should be initiated to encourage people to consume more fruits and vegetables (as well as whole grains and beans). Currently, the "5 A Day" program of the U.S. Department of Health and Human Services, which encourages people to eat more fruits and vegetables, receives only about $5 million in annual funding and has negligible impact. Other media campaigns should discourage the consumption of fatty meat and dairy products, soft drinks, and salty processed foods. The overall budget should be at least $150 million per year (50 cents per person).
- The U.S. Department of Agriculture's (USDA's) highly successful Fruit and Vegetable Snack Program provides a free serving each day of a fruit or vegetable to schoolchildren. Unfortunately, the program only has the funding to reach several hundred schools. Considering that it benefits both children and farmers, it should be expanded nationally at an annual cost of roughly $4 billion. That would be a far smarter investment than the $20 billion paid in some years to grain, cotton, and rice farmers.

- In the Food Stamp program, bonus stamps could be provided for the purchase of fresh, frozen, canned, or dried fruits and vegetables. Similarly, as the Institute of Medicine has recommended, the Women, Infants, and Children (WIC) program should provide more fresh fruits and vegetables and less juice, cheese, milk, and eggs. The USDA is required to rewrite its regulations by November 2006.

- City and state governments should sponsor more farmers' markets, especially in low-income communities, to help distribute locally grown fresh produce.

2. Reduce the Fat Content of Meat

Because animal fat promotes heart disease, it would be helpful if beef and pork were as lean as possible. (Hog farmers are raising far leaner hogs than they did several decades ago.) Certain breeds of cattle tend to be lower in fat, and younger animals usually are lower in fat than older ones. Fat content is also increased by the high-grain diets cattle eat at feedlots. The government should implement policies that lower the fat content of meat products.

- The approximate fat content of cattle could be assessed at the slaughterhouse, with a modest per-pound tax levied on higher-fat cattle. The revenues from that tax could be used to reward ranchers and feedlot operators who deliver lower-fat cattle to market, encourage farmers to raise lower-fat breeds and feed cattle grain for shorter periods of time, and encourage consumers to choose lower-fat and pasture-raised beef products. Though pigs are slimmer than ever, analogous programs could ensure that that healthy trend continues.
- The USDA has standards of identity that limit the fat content of certain processed meats, but the current limits are a generous 30 percent by weight in ground beef and hot dogs and 50 percent in pork sausages. (The average hot dog now contains more than twice as much fat as protein.) Those high-fat products provide a ready market for fat trimmings from cattle and hogs and clog consumer arteries. The fat limit for ground beef and hot dogs should be lowered—perhaps over several years—to 20 percent and for pork sausages to 25 percent. Judging from the many lower-fat products already on the market, companies could lower the fat content and still provide good-tasting foods.

3. Reduce the Fat Content of Milk

The saturated fat in cow's milk is a leading cause of heart disease. Although individuals now can choose lower-fat dairy products, the fat that is removed inevitably returns to the market in the form of butter, cream, ice cream, or other high-fat products. To reduce the volume of saturated fat entering the food supply, dairy pricing policies should be revised to encourage farmers to deliver lower-fat milk. Producers that deliver milk lower in saturated fat could be paid more. The money could come—in a zero-sum manner—from lower payments for milk that is higher in saturated fat. Several approaches can improve the nutrient content of milk.

- Use breeds of cows that provide milk with less total and saturated fat, and, within those breeds, select for propagation individual cows whose milk is lower in fat.
- Add conjugated linoleic acid to the cows' feed or change feed in other ways to lower the *total* fat content of milk by about 25 percent.[2]
- Add canola seeds (the source of canola oil) or other sources of unsaturated oil to cows' feed to lower the *saturated fat* and increase the *unsaturated fat* content by 20 percent each.[3]

"Spreadable" butter is produced naturally in Ireland by feeding cows a source of unsaturated fat, such as canola.

4. Label Food More Effectively

The familiar Nutrition Facts label on packaged foods is used daily by millions of people, but it has not been as effective as some had hoped in improving diets and promoting health; also, fish, produce, and unprocessed meat and poultry are not required to have nutrition labels.

- The Food and Drug Administration (FDA) and the USDA should develop a more effective labeling system to supplement the Nutrition Facts label. One option would be to require companies to put a symbol on the front of a product's package to highlight the food's overall nutritional value. Foods would be rated according to their content of saturated

Swedish voluntary "good food" symbol.

fat, sodium, vitamins, and other nutrients and then required to put a green ("any time"), yellow ("sometimes"), or red ("seldom") circle or square on the front label. An alternative more palatable to industry would be to establish a voluntary system, as the United Kingdom and Sweden have done (see image). Such straightforward front-label symbols would be a great help to hurried shoppers, children, and others.

- Steaks, ground beef and poultry, and other fresh and frozen meat and poultry products are not required to provide nutrition information on labels. The USDA should order such labeling, which would help people avoid fattier foods.
- While nutrition information is on most packaged foods, people choose blindly when they eat out. Chain table-service restaurants should be required to list on their menus the calorie, saturated and trans fat, and

sodium content of each item. Fast-food restaurants should be required to post the calorie content of each item on their menu boards.[4]

5. Prevent Foodborne Diseases

Animals harbor a wide variety of microorganisms that do not harm the animals but can cause serious and sometimes fatal diseases in humans. Farming and processing practices, as well as the federal government's regulatory system, should be improved to minimize the toll of foodborne diseases. Aside from eating (and producing) less meat, actions to prevent food poisoning include the following.

- The health and food-safety responsibilities of the USDA, FDA, and other federal agencies should be consolidated into a single independent agency, as several other countries have done. The current multi-agency system is inefficient and suffers from a severe conflict of interest: The USDA is charged with both promoting the consumption, and regulating the safety, of meat and poultry products. A new, streamlined public health agency should be empowered to levy stiff fines, recall tainted products from the marketplace, and inspect foreign processing plants.
- Congress should give the federal government a specific mandate to reduce hazards in the food supply. Pathogen-reduction and enforcement authority are largely lacking from our existing meat and poultry inspection laws.
- The USDA should require cattle ranchers to use bar-coded or radio-frequency identification tags or retinal imaging to track individual cattle from birth to the slaughterhouse and to help pinpoint sources of food poisoning and mad cow disease. Such systems are already in use in Europe, Canada, Japan, and other countries. Fresh produce and grains should carry information on the country (and possibly the state and farm) of origin to facilitate traceback in the event of contamination.
- Food-safety measures—from the farm to the supermarket—should be upgraded. Vaccinations, feed additives, carcass washes, temperature controls on trucks, and other measures are needed to minimize the presence of pathogens. On egg farms, for example, layer hens should be certified as *Salmonella*-free, and any eggs that might be contaminated with *Salmonella* should be pasteurized or cooked in processed foods.
- Outbreak reporting systems should be improved to encourage more thorough investigations and more specific information on the food sources of the outbreaks. When animal pathogens are found in plant-based foods, investigators should identify how and where the contamination likely occurred.

6. Prevent Antibiotic Resistance

Many farmers add medically important antibiotics to livestock feed to compensate for overcrowded and unsanitary conditions. That practice, however, increases the likelihood that bacteria harmful to humans will become resistant to antibiotics and cause infections that are more difficult to treat. To maintain the effectiveness of those invaluable drugs for human medicine, Congress should ban the routine feeding of medically important antibiotics to livestock. Indeed, scientific research and a few major producers have found that feeding antibiotics to healthy chickens, hogs over 50 pounds, and grass-fed cattle is largely unnecessary. A ban, as likely would occur if pending bipartisan legislation (S.742) were to pass, would not prevent farmers and ranchers from using antibiotics to treat sick animals.

7. Stop Promoting Unhealthy Meat and Dairy Foods

The beef, pork, dairy, and egg industries, with administrative assistance from the USDA, "tax" themselves to raise war chests for advertising (for example, "Pork–The Other White Meat," celebrity "milk mustache" ads, "Beef Gives Strength," "The Incredible Edible Egg") and research.

Together, those industries spend tens of millions of dollars annually promoting their products—a sum that dwarfs what the government and industry spend to promote the consumption of fruits, vegetables, and whole grains. Although some of the advertising features lower-fat types of meat and milk, all of it serves as an advertisement for foods that contribute to health and environmental problems. It is hypocritical for the government to facilitate the promotion of foods that are inconsistent with its own dietary guidelines. Congress should eliminate federal involvement in the milk, cheese, beef, pork, and egg programs or limit the advertising to the healthiest products.

8. More Healthful Meals at Government-Run Facilities

Federal, state, and local governments directly feed millions of people every day at schools, government cafeterias, military bases, prisons, and hospitals. Government could easily promote healthier diets at those facilities by providing more dishes based on fruits, vegetables, beans, and whole grains. Animal products served should be low in fat and salt and made from animals raised humanely and without medically important antibiotics. Nutrition information should be provided. Such government efforts would promote health; create markets for healthier foods; and set an example for other large employers, hospitals, colleges, and restaurants.

Improving the Environment

A nation's stewardship of the environment reflects its consideration of future generations. Today, though, farmers apply copious amounts of fertilizer and pesticides to vast acreages of crops destined for animal feed, polluting the environment and possibly harming wildlife, farmworkers, and consumers. Raising large numbers of cattle, hogs, and poultry in concentrated animal feeding operations (CAFOs) generates air and water pollution.

Numerous state and federal laws are aimed at protecting the environment, but some of those laws have limited applicability to agriculture. Moreover, in its regulation of the industry, the federal government sometimes has gone in the wrong direction: The Environmental Protection Agency (EPA) has exempted some 14,000 poultry, egg, dairy, and hog farms from potential fines of up to $27,500 per day for polluting the air or water with animal manure.[5]

To tackle the noxious problems caused by CAFOs, local and national citizens' groups, including Public Citizen and Global Resource Action Center for the Environment, are seeking to stop the building of new large animal feeding operations. In 2003 the American Public Health Association joined in, urging federal, state, and local governments to impose a moratorium on new CAFOs until adequate scientific data on the "risks to public health have been collected and uncertainties resolved."[6] Counties in Iowa, Missouri, North Carolina, and other states where the hog industry has been most aggressive are beginning to approve moratoriums on CAFOs.

Shifting to a more plant-based diet is one sure way to lessen numerous environmental burdens. But since not everyone is going to do that, federal and state governments should adopt new policies to protect the environment from large-scale animal agriculture. The following measures also would nudge people in a more plant-based direction by slightly increasing the costs of producing beef, pork, poultry, eggs, and milk. After all, it is only fair that livestock producers—and consumers of animal products—bear the full economic costs of their activities. Even the Farm Foundation, which is supported by the cattle, hog, and other industries, acknowledges that "reflecting the true cost and value of manure and byproducts in prices of products or services might provide an incentive for producers and processors to adopt systems that maximize profits while being environmentally friendly."[7]

1. Prevent Air Pollution from Factory Farms

Factory farms that raise cattle, hogs, and poultry are major air polluters. Governments should limit the density and total number of animals. The EPA should aggressively enforce the Clean Air Act, Superfund (a waste abatement program), and Community Right-to-Know laws as they apply to CAFOs.

2. Prevent Water Pollution from Factory Farms

In its place, nutrient-rich manure is a valuable resource. But the 1 trillion pounds of animal waste generated by animal feeding operations frequently pollutes nearby streams and rivers.[8] When manure lagoons on hog farms are breached—because of major storms, equipment breakdowns, or operational errors—the waste pollutes groundwater and nearby waterways, contaminating the water and killing fish. In addition, nutrients in animal manure applied to cropland often pollute waterways.

● Water pollution would be best mitigated by raising fewer animals and limiting the size of CAFOs. Short of that, the EPA has mandated that CAFOs, as well as smaller or less-intensive feeding operations likely to cause water pollution, obtain permits to limit pollution. Those permits include comprehensive nutrient management plans, the requirements of which are designed by the USDA and vary by state. Management plans are not now, but should be, subject to public review to promote enforcement. The plans also do little to stop the construction or operation of open-air lagoons of dirty, smelly manure. Stringent Clean Water Act permits with enforceable provisions should be used to prevent pollution from CAFOs' manure storage facilities and when the manure is spread or sprayed on fields. Considering how troublesome manure lagoons have been, the EPA could ban ones over a certain size.

- The USDA's Environmental Quality Incentives Program (EQIP) gives individual CAFOs up to $450,000 to cover the cost of building, improving, or upgrading their manure lagoons and effluent sprayfields. However, as the *New York Times* put it, that largesse helps farmers "comply with regulations that don't mean much to begin with."[9] EQIP grants encourage the use of large-scale lagoons and sprayfield systems, because they do not limit the size of the operations that receive the grants. EQIP support should be limited to smaller, less environmentally harmful livestock facilities.

- To prevent phosphorus pollution of waterways, the USDA should encourage farmers to use more appropriate animal feed. Two common approaches are reducing phosphorus levels in feed and adding the enzyme phytase, which breaks down the phosphorus-rich phytic acid in feed and enables animals to absorb more phosphorus. Adding phytase to swine and poultry feed reduces the phosphorus content of manure by as much as 25 to 50 percent.[10] Dairy farmers, who commonly add too much phosphorus (and nitrogen) to feed, could reduce costs and runoff substantially if they cut back.[11]

- Government agencies generally give broad discretion to producers to create and implement nutrient management plans, though some states might impose more stringent requirements. Only Wisconsin has allowed nutrient feed-management changes to qualify for funding from the USDA's EQIP.[12] Other states should do the same.

- Hormones—including natural ones and the growth hormones implanted in beef cattle—have been found in waterways downstream from feedlots. Initial studies show that the minuscule amounts of hormones cause malformations in fish. Greatly decreasing the number and density of cattle in CAFOs would solve the environmental problem.

3. Reduce Water Use

The enormous amounts of (mostly irrigation) water that are used to produce feed grains erode soil, pollute water, deplete groundwater reservoirs, and poison fish. Ultimately, over-irrigation can deplete water of oxygen and harm wildlife in and around ponds and lakes.

- Irrigation subsidies encourage farmers to waste water and cultivate poor-quality land where irrigation contributes to water-quality problems. From 1902 to 1986, irrigation subsidies cost taxpayers as much as $70 billion. The subsidies just to farmers in California's Central Valley Project now amount to $400 million a year, mostly going to large farmers, according to the Environmental Work-

ing Group.[13] Those subsidies should be reduced or eliminated. Currently, many water rate structures charge farmers on a per-acre basis regardless of water use. Water deliveries to farms should be measured and farmers charged according to how much water they use.
- Federal loans or grants should be available to encourage farmers to use more efficient irrigation systems. A portion of farm subsidies could be withheld from farmers who waste water.

4. Reduce Pesticide and Fertilizer Use

Gargantuan quantities of fertilizer and smaller quantities of pesticides help maximize yields of feed grains and other crops but exact a cost from the environment and health. The mining of minerals and manufacture of fertilizer require huge amounts of energy and nonrenewable resources and pollute the air and water. Using the fertilizer generates more air and water pollution. Pesticides may harm workers and nearby residents, as well as non-target animals and plants. And consumers, of course, would prefer not to have pesticide residues in their food. Eating fewer animal products would reduce the harm from pesticides and fertilizer (though the benefits would be slightly reduced because more food crops would have to be produced). Government actions to lessen the problems include the following.

- The USDA and state departments of agriculture should mount intensive programs to encourage feed-grain producers (and other farmers) to slash their use of pesticides and chemical fertilizer by using techniques ranging from integrated pest management to organic farming to biotechnology. Though agriculture departments have long belittled organic agriculture, they are beginning to see that thousands of small farmers are thriving by growing fruits, vegetables, grains, and livestock for that exploding niche market. Just as the European Union provides about $500 million a year in subsidies to organize farmers, states could provide loans, grants, or tax breaks (see next item) to help farmers get off the chemical treadmill.[14]

• Taxing fertilizers and pesticides would internalize some of their environmental costs and reduce their use. Even a small tax, which would

not affect food prices, would raise significant revenues to fund research projects and support improved farming practices. But currently, many farm states actually exempt pesticides and fertilizers from sales taxes, at a cost to the states of hundreds of millions of dollars each year.[15] The Soil and Water Conservation Society estimates that a 5 percent federal tax on agricultural fertilizers and chemicals could raise $1 billion annually.[16] The key is to earmark tax revenues for environmen-

Insecticides all too often kill harmless and helpful insects, such as ladybugs, along with pests.

tal and health programs. Nebraska uses pesticide registration fees ($1.3 million in 2003) to fund conservation programs, including installation of conservation buffers, weed control, and water quality improvements. Iowa's Groundwater Protection Act taxes nitrogen fertilizer (75 cents per ton) and imposes pesticide registration fees to support conservation activities. In 2001, the state's fertilizer tax raised $913,000, and its pesticide fees raised $2.7 million, with 35 percent of the revenue allocated to

the Leopold Center for Sustainable Agriculture and the remainder to solid waste and agricultural health programs.[17]

• Even though the 2002 Farm Bill requires producers to reduce soil erosion (through the USDA's Conserva-

tion Compliance provisions) and protect wetlands (under the Swamp-buster provisions) in order to receive subsidies, no such requirement directly protects water quality. Subsidies could be made contingent upon farmers' reducing fertilizer and pesticide inputs to appropriate levels.

- The Netherlands, in response to a European Union directive aimed at protecting the environment, has tested several approaches to limiting nitrogen and phosphorus from fertilizer and manure. It recently implemented a complex system of application limits. The USDA could follow that example and seek congressional authority to sponsor pilot projects that would limit fertilizer and manure use.[18]

5. Reduce Feed Grain Usage

America's livestock industry relies heavily on feeding practices designed to bring meat and dairy products to market as quickly and cheaply as possible.

The effects of those feeding practices on the environment, the animals, and the nutrient content of foods have received scant attention. Reducing the amount of grain in cattle feed (and allowing chickens and pigs to obtain at least some of their food from barnyards and pastures) would deliver healthier products to consumers, protect the environment, and protect the animals' welfare. Consumers could help move the country in that direction by eating fewer animal products or, at the very least, choosing grass-fed beef.

- Congress should provide greater funding for programs that pay feed-grain farmers to remove large areas of land, especially environmentally sensitive land, from production. Slightly higher grain prices might slightly reduce the amount of grain fed to cattle.
- The high-grain diets fed to cattle at feedlots makes the cattle sick; increases the fat content of the beef; and necessitates the use of fertilizer, water, pesticides, and land to produce the grain. To protect people, cattle, and the environment, the USDA or FDA should set standards that would limit the grain content of the feed and the length of time cattle eat it.
- The USDA should develop a labeling system that would identify meat, poultry, and milk produced in an environmentally friendly and humane

Cheap Corn: Indirect Subsidy to Livestock Producers

Since the Depression, the federal government has maintained farm programs to keep farmers afloat and supplies and prices stable. The traditional policies were changed radically in 1996, when Congress passed the Freedom to Farm legislation. This simplified description of farm programs highlights some key points.[19]

Before 1996, corn and several other major crops had price floors. The government used a combination of crop-storage programs and acreage limitations to support the prices for those crops. Grain merchants such as Cargill and food processors such as General Mills, as well as foreign buyers, had to pay at least the floor prices for grains. The 1996 law ended most planting restrictions and replaced price supports with government payments.

One subsidy consists of direct payments to farmers regardless of the amount of crops produced. The original goal was to phase out those payments over several years.

A second subsidy, using so-called loan deficiency payments (the loans actually are not intended to be repaid), pays farmers the difference between the market price and the "support price" for corn, wheat, cotton, and other "program crops." Before 1996, if the support price, also called the loan rate, was $2.00 per bushel, farmers were essentially assured that they would receive at least $2.00 per bushel for their grain. Now, with the use of loan deficiency payments, the farmer receives the market price, for example $1.50 per bushel, plus the 50-cent difference from the government. Those purchasing grain pay $1.50 per bushel (not $2.00), which is well below the cost of producing the grain. That is, the purchasers of the grain buy the grain at a subsidized price.

That system worked well while market prices were high, but when prices dipped a couple of years after the 1996 farm bill was passed, Congress began providing farmers with emergency subsidies—billions and billions of dollars' worth of subsidies. Without a program to reduce production, such as acreage set-asides, the government did not possess any policy tools to boost crop prices. The 2002 farm legislation replaced ad hoc emergency payments with a third subsidy that kicks in when prices are low.

That set of three subsidies helps farmers when harvests are large and prices are low. Between 1995 and 2004, according to the Environmental Working Group, corn growers "farmed the government" for $42 billion; subsidies for all crops totaled $144 billion. In 2005 alone, farm subsidies totaled $23 billion. Importantly, those subsidies constitute a multibillion-dollar-a-year boon not just to farmers, but also to livestock growers, food processors, and exporters.

One solution to costly subsidies would be to return to price floors. That would ensure that buyers paid a price that reflected the actual direct costs of growing corn (though not the costs of pollution). The higher price of animal feed would encourage the cattle industry to reduce the time cattle spent at feedlots and might increase slightly the cost of beef and, somewhat more so, the cost of pork and chicken.

To keep the price of corn up near the target price, production might have to be kept down by limiting the acreage planted in corn, expanding programs that protected environmentally sensitive land, and designating acreage that had to be planted in crops, such as switchgrass, that could be burned or converted to ethanol for cost-efficient energy. In addition, the government should, as it used to do, ensure reserves of enough wheat, corn, and other crops to protect against droughts or other calamities here or abroad.

Some of the billions of dollars saved by changing farm subsidy programs (including limiting payments to large farmers) should be reinvested in the farming community and food policies. Programs should help farmers reduce their use of fertilizer and pesticides, transition to organic methods, and raise pasture-fed cattle. Smaller farmers, especially ones within driving distance of major cities, should be helped to sell produce directly to supermarkets, schools, and consumers at farmers' markets. The tiny Fruit and Vegetable Snack Program that provides free daily snacks to children in a couple of hundred schools should be greatly expanded.

manner. (In 2006, the USDA took a step in that direction by proposing a definition for "grass-fed" cattle and sheep.)

6. Prevent Overgrazing on Public Lands

Overgrazing of cattle on public lands contributes to riparian damage (that is, damage to surrounding waterways), erosion and water pollution, and harm to endangered species. Ranchers who use public lands in the West save as much as $500 million a year because the federal government absorbs most of the costs of managing the land.[20] Grazing fees should be increased to reflect the true cost to the government.

Another approach would be to buy out ranchers' grazing rights through a voluntary program. A 2003 federal buy-out bill—the Voluntary Grazing Buyout Act—had strong support from environmental, conservation, and animal welfare organizations, and some ranchers, but died because of opposition from the National Cattlemen's Beef Association.

Improving Animal Welfare

How a nation treats farm animals is a good gauge of that nation's compassion. Most animals raised on contemporary factory farms live in tiny spaces; breathe foul air; wallow in their own manure; eat unnatural diets;

U.K. Farm Animal Welfare Council's
5 Freedoms[23]

The welfare of an animal includes its physical and mental state and we consider that good animal welfare implies both fitness and a sense of well-being. Any animal kept by man must, at least, be protected from unnecessary suffering....

1. **Freedom from Hunger and Thirst**—by ready access to fresh water and a diet to maintain full health and vigour.

2. **Freedom from Discomfort**—by providing an appropriate environment including shelter and a comfortable resting area.

3. **Freedom from Pain, Injury or Disease**—by prevention or rapid diagnosis and treatment.

4. **Freedom to Express Normal Behaviour**—by providing sufficient space, proper facilities and company of the animal's own kind.

5. **Freedom from Fear and Distress**—by ensuring conditions and treatment which avoid mental suffering.

or endure branding, castration, debeaking, or de-tailing. In some cases, animals are intentionally harmed—or even tortured—by workers.

New laws must be adopted and vigorously enforced to ensure that all animals are raised and handled humanely, from birth to slaughter (see "U.K. Farm Animal Welfare Council's 5 Freedoms," above). Recent laws in California, to prohibit the force-feeding of ducks and geese for foie gras, and in Florida, banning the housing of pregnant sows in cramped crates, demonstrate broad public support for protecting farm animals. Overseas, sow gestation crates have been banned in the United Kingdom and are being phased out in the European Union and New Zealand. Similarly, layer hen battery cages were banned in Switzerland 10 years ago and are being phased out in the European Union.[21]

Reforms aimed at farm practices and to encourage consumer purchase of foods made from humanely raised animals certainly would improve animal welfare. Matthew Scully, author of *Dominion: The Power of Man, the Suffering of Animals, and the Call to Mercy*, a book pleading for more humane treatment of animals, proposed a federal Humane Farming Act that "would explicitly recognize animals as sentient beings and not as mere commodities or merchandise."[22] The problems discussed in Argument #6 ("Less Animal Suffering") suggest that such a law should:

● Impose ample space requirements to prevent crowding of farm animals and eliminate restrictive caging of hogs, layer hens, and veal calves.

- Regulate conditions such as temperature, water, and space per animal on the trucks and railroad cars that transport animals from farm to farm or to feedlots and the slaughterhouse.
- Slow slaughterhouse lines to help ensure that the animals are stunned properly.
- Ensure that slaughterhouse operators and workers abide by the federal Humane Methods of Slaughter Act, provide for improved enforcement of that law by the USDA, and extend the law to include poultry.
- Establish a reliable labeling scheme to encourage consumers to buy meat, dairy products, and eggs from more-humanely raised animals and to inform consumers when animals are raised on factory farms.
- Limit the amount of grain in feed and the duration of grain feeding at cattle feedlots. Doing that also would lessen the need for grain and antibiotics, reduce pollution from feedlots, and probably lead to lower-fat beef.
- Require cattle to be identified with ear tags or other devices. That would mitigate the need for hot-iron branding and also help health officials identify animals infected with dangerous bacteria or the prions that cause mad cow disease.
- Require husbandry and cleanliness standards to reduce use of antibiotics.

States, too, should enforce their animal cruelty laws as they apply to farm animals or amend laws that exempt farm animals from protection. In the absence of federal action, states should prohibit practices such as debeaking and forced molting of chickens, hot-iron branding of cattle, and de-tailing of hogs.

Such measures would modestly raise the price of animal products, but any society that considers itself civilized should ensure that farm animals are treated humanely.

Appendixes and Notes

Appendix A.
A Bestiary of Foodborne Pathogens
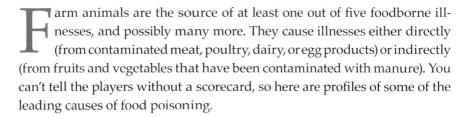

F arm animals are the source of at least one out of five foodborne illnesses, and possibly many more. They cause illnesses either directly (from contaminated meat, poultry, dairy, or egg products) or indirectly (from fruits and vegetables that have been contaminated with manure). You can't tell the players without a scorecard, so here are profiles of some of the leading causes of food poisoning.

Campylobacter jejuni

As the leading cause of foodborne illness in the United States, *Campylobacter* causes 2.4 million sicknesses and over 100 deaths each year.[1] The main symptom of campylobacteriosis is diarrhea that lasts up to 10 days; it recurs in one out of four people.[2] In severe cases, people can die from septicemia (a bacterial infection in the blood) or hemolytic uremic syndrome (a cause of short-term kidney failure in children, usually following an infection in the digestive system). Longer-term effects of some infections include arthritis; meningitis (an inflammation of the central nervous system); colitis, which results in ulcers in the large intestine; and cholecystitis (inflammation of the gallbladder). Each year, several thousand people who had contracted campylobacteriosis later develop Guillain-Barré

Syndrome, an autoimmune disorder that causes severe weakness and even paralysis.[3]

Campylobacter in feces, water, and urine remains viable at refrigerator temperatures for several weeks.[4] It thrives on the manure-strewn floors of factory farms, and the animals living there are regularly infected.

Campylobacter is mainly found in poultry and sometimes in cattle. Healthy chickens and turkeys can carry the bacteria, which spread easily through flocks. A 1999 U.S. Department of Agriculture (USDA) study found *Campylobacter* in over 90 percent of poultry.[5] During slaughter, bacteria in the intestines may contaminate the meat, so it was not surprising that researchers at the Centers for Disease Control and Prevention (CDC) found that 44 percent of chickens at supermarkets were contaminated with *Campylobacter*.[6] Even more alarming, 24 percent of the *Campylobacter* isolates were resistant to the powerful antibiotics—fluoroquinolones—that are often a last resort for treating food-poisoning victims. (Those antibiotics are no longer allowed to treat poultry flocks; see "Factory Farming's Antibiotic Crutch," p. 68.) Despite all the concerns about foodborne illnesses, illness rates showed little change between 1999 and 2004.[7]

Clostridium perfringens

About 250,000 cases of food poisoning—but only a handful of deaths—are caused each year by *Clostridium perfringens*. That bacterium is normally found in the intestinal tracts of cattle, hogs, poultry, and fish, as well as in humans, and it is widely distributed in soil. Because *C. perfringens* produces spores that can survive in boiling water, it is often present after cooking, and bacterial populations may increase while foods cool. Fully cooked meat and gravy are the most common causes of infections. Symptoms include severe abdominal cramps and diarrhea lasting for up to two weeks. Rarely, the infection may progress to the potentially fatal pig-bel syndrome, which destroys the intestines.[8]

Escherichia coli

E. coli is a natural and abundant resident of the intestinal tract of humans and most other mammals. In a healthy person, the bacteria can help prevent disease—particularly foodborne illnesses—by out-competing pathogens for nutrients. But certain subspecies of *E. coli*—particularly *E. coli* O157: H7—cause gruesome foodborne illnesses.

E. coli O157:H7 produces a toxin that damages the lining of the intestine, causing bloody diarrhea. The infection may lead to kidney failure and death. Each year, *E. coli* O157:H7 and its close relatives cause roughly 94,000 illnesses and kill about 80 people.[9] Infections tend to be most severe in children. That was the case in 1993 when contaminated beef patties served at Jack in the Box restaurants were linked to more than 600 illnesses and the deaths of four young children.[10]

E. coli O157:H7 resides in up to 6 percent of cattle and one-third of sheep. The bug typically infects humans through meat contaminated at slaughterhouses. During evisceration, workers may damage the intestine, releasing its bacteria-laden contents onto meat.[11] The nasty germ also can spread from cattle hides contaminated with manure to slaughterhouse workers and equipment—and then to meat.[12] Thanks in part to the beef industry's vigorous efforts to clean up its operations, illnesses caused by E. coli O157:H7 declined by 42 percent between 2002 and 2004.[13] Still, in 2003, 60 percent of the nation's largest meat plants failed to abide by federal regulations.[14]

These germs can infect people through routes other than food. After a group of schoolchildren visited a Pennsylvania dairy farm, 51 children were sickened by a strain of E. coli O157:H7 identical to that found in cattle on the farm.[15] And at a county fair in New York, runoff from a dairy barn contaminated the unchlorinated water supply, sickening 1,000 people with E. coli O157:H7 and causing two deaths.[16]

Listeria

Listeria monocytogenes is one of the deadliest foodborne pathogens. The CDC estimates that *Listeria* causes 2,500 flu-like illnesses each year—20 percent of which are fatal.[17] The latest data indicate that illness rates stayed the same between 2000 and 2004.[18]

Listeria is widely distributed in the environment and is a hardy bacterium that can survive freezing, drying, and heat remarkably well. And, increasing its threat, *Listeria* can grow at refrigerator temperatures.[19]

Livestock may get infected by eating contaminated feed.[20] Most human infections result from tainted meat, though vegetables can be contaminated by *Listeria* in the soil, irrigation water, or manure used as fertilizer.[21]

Listeria is particularly harmful to people who are immunosuppressed, such as the elderly, those with organ transplants, and people with HIV. At highest risk, however, are pregnant women. Because of hormonal effects on the immune system during pregnancy, pregnant women are 20 times more likely to contract a *Listeria* infection than other people and account for about one-third of all cases.[22] *Listeria* can cross the placental barrier and infect the developing fetus, which often results in miscarriages and stillbirths. It can also cause meningitis, which, if the child survives, may result in cerebral palsy or other chronic neurological illnesses. Pregnant women should not eat the foods most likely to be contaminated: hot dogs, deli meats, soft cheeses, paté, and smoked seafood.[23]

Mad Cow Disease

Mad cow disease, or bovine spongiform encephalopathy (BSE) infects the brains of cattle, resulting in a cerebrum so riddled with holes that it resembles a sponge. Infected cattle have trouble standing and walking, hence the term "mad." As of April 2006, only eight cases of mad cow disease have been confirmed in North America—five in Canada and three in the United States—but the human form of the disease (variant Creutzfeldt-Jakob disease, or vCJD) is so horrible that it has received enormous attention in the media, and deserved attention from government and industry.

People can suffer vCJD by eating contaminated beef. The infectious agents—called prions (improperly folded proteins)—cause brain decay similar to that seen in cattle.[24] Victims suffer memory loss, impaired coordination, and hallucinations, and they invariably die. In the United Kingdom, the epicenter of mad cow disease, 3.7 million cattle were slaughtered to stop its spread.[25] As of May 2006, 155 Britons had died from the disease (and fewer than 20 in other countries).[26]

Several factors contribute to prions' harmfulness. First, prions have a disturbing ability to jump from one species to another, including cattle, humans, sheep, and domestic cats.[27] Also, prions are far hardier than bacteria and viruses. They can withstand boiling, the even higher temperatures used for sterilizing medical equipment, freezing, irradiation, and most acids and bases.[28]

BSE is believed to have spread by adding rendered leftover meat and bones of cattle carcasses to cattle feed as a source of protein.[29] A single infected cow could infect a large and geographically diverse population of other cows.

The risk to humans of consuming infectious meat is increased by some processors' use of advanced meat recovery (AMR). That mechanical process extracts hard-to-reach bits of meat from the bones after most of the meat has been removed by hand. Because spinal columns are often processed in AMR equipment, the resulting meat may contain high-risk spinal and central nervous system tissue. In 2003, the USDA found such tissue in meat from more than 75 percent of the plants using AMR equipment. Though AMR use is declining, millions of pounds of AMR-recovered beef are used each year in such products as ground beef, meatballs, and taco filling.[30]

Fortunately, the chances of contracting vCJD in the United States are infinitesimal, because so few cattle are infected. Increasingly stringent controls are reducing the risk even further. Only one confirmed case has occurred in the United States, and that involved a person who spent time in the United Kingdom when mad cow disease was at its height. That also was true of most of the infected individuals in other countries.[31] Eating beef is far likelier to cause heart disease than vCJD.

Salmonella

A *Salmonella* infection can be contracted only by eating contaminated food.[32] Infections cause flu-like symptoms—including vomiting, diarrhea, and fever—in over 1 million Americans each year. *Salmonella* kills about 600 Americans each year, with its victims generally being 65 or older.[33] Rarely, salmonellosis causes Reiter's syndrome, a form of arthritis.[34] Illness rates have barely changed over the past 10 years.[35]

Salmonella can live in the digestive tracts of most vertebrates, including cattle, hogs, and poultry. It can inhabit the ovaries of laying hens, contaminating their eggs before the shells form. On hog farms, *Salmonella* can survive for months in manure slurry.[36]

The USDA found *Salmonella* in over half of all large feedlots, 5 percent of dairy cows on farms, and 15 percent of the dairy cows sold at livestock auctions. The USDA also found *Salmonella* in 9 percent of broiler chickens and 38 percent of hog operations.[37]

Antibiotic resistance increases the harmfulness of *Salmonella* infections, with some strains being resistant to several different antibiotics. A study by the Food and Drug Administration and the University of Maryland found that 20 percent of samples of ground chicken, beef, turkey, and pork were contaminated with *Salmonella*, and 84 percent of those bacteria were resistant to at least one antibiotic.[38]

In recent years, the biggest *Salmonella* problem has been in eggs.[39] About 1 out of every 20,000 eggs is contaminated with *Salmonella*.[40] Assuming that half the eggs are eaten fresh, Americans have about a 1 in 75 chance of being exposed to *Salmonella* through eggs over the course of a year. Proper handling and cooking can usually kill the germs.

The practice of forced molting of layer hens accelerates the spread of *Salmonella*. When they are deprived of food, water, and light to prolong their productive lives, molted hens are 100 to 1,000 times more susceptible to *Salmonella* infection than unmolted birds.[41]

Staphylococcus

Over 185,000 cases of food poisoning each year are caused by *Staphylococcus*. Meat, poultry, and dairy products are the most common food causes of those infections. Because *Staphylococcus* is often present on human skin, poor sanitation among food handlers is probably the primary problem. However, animals also carry *Staphylococcus*, and this bug has been found in the air around hog barns and on the hides of most livestock.[42]

Staphylococcus's toxin may be produced in food before cooking, and it is stable at high temperatures. Symptoms of infection include nausea, vomiting, headache, muscle cramps, and fluctuations in blood pressure and pulse rate. Symptoms usually last about two days.[43]

Toxoplasma gondii

Toxoplasma gondii is a parasite that causes about 115,000 foodborne illnesses each year, killing 375 people. Toxoplasmosis occurs when *T. gondii* infects people who eat undercooked pork, lamb, or other meats or who improperly handle those foods during preparation. The infection can also occur through exposure to cat feces. Minor infections resemble the flu, but *T. gondii* also can enter the central nervous system, damaging the eyes or brain.[44] In pregnant women, infections can cause miscarriages or birth defects.[45]

Appendix B.
Eating Green Internet Resources

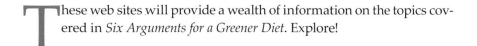

These web sites will provide a wealth of information on the topics covered in *Six Arguments for a Greener Diet*. Explore!

- Center for Science in the Public   www.cspinet.org
  Interest   www.eatinggreen.org

Agriculture

- U.S. Department of Agriculture   www.usda.gov
  (Agricultural Research Service,
  Economic Research Service, food
  consumption data, etc.)

- GRACE Factory Farm Project   www.factoryfarm.org
  — Web animated movie   www.themeatrix2.com

- Sierra Club (factory farms)   www.sierraclub.org/factoryfarms/

Animal welfare

- Compassion in World Farming   www.ciwf.org.uk/
- Compassion Over Killing   www.cok.net
- Farm Sanctuary   www.farmsanctuary.org; www.factoryfarming.com
- Humane Society of the United States   www.hsus.org
- People for the Ethical Treatment of Animals' vegetarian campaign   www.GoVeg.com; www.Meat.org (video)

Antibiotics

- Alliance for the Prudent Use of Antibiotics   www.apua.org
- Keep Antibiotics Working   www.keepantibioticsworking.org

Environment

- Environmental Defense (global warming, air and water pollution, toxic chemicals)   www.environmentaldefense.org
- Environmental Working Group (farm subsidies, chemical contaminants)   www.ewg.org
- Monterey Bay Aquarium's Seafood Watch (choosing safe and abundant seafood)   www.mbayaq.org/cr/seafoodwatch.asp
- Natural Resources Defense Council (pesticides, land use, global warming)   www.nrdc.org
- U.S. Environmental Protection Agency (pesticides, air and water pollution, etc.)   www.epa.gov

Nutrition and health

- American Institute for Cancer Research (diet and cancer; recipes)   www.aicr.org
- Centers for Disease Control
  - NHANES dietary intake studies   www.cdc.gov/nchs/nhanes.htm
  - healthy recipes   www.cdc.gov/nccdphp/dnpa/5aday/recipes/index.htm

- 5 A Day (produce industry's web site, with recipes)   www.5aday.com

- Harvard School of Public Health's "Nutrition Source" (reliable, independent source of information)   www.hsph.harvard.edu/nutritionsource/

- National Institutes of Health
  - Medline Plus (sensible, mainstream information)   www.nlm.nih.gov/medlineplus/
  - PubMed (abstracts of medical research)   www.pubmed.gov

- U.S. Department of Agriculture resources (*Dietary Guidelines for Americans*, Food Guide Pyramid, food safety)   www.nal.usda.gov/fnic/
 www.nutrition.gov
  - Nutritional value of foods   www.nal.usda.gov/fnic/foodcomp/Data/HG72/hg72_2002.pdf
  - Food and nutrient consumption   www.ers.usda.gov/Data/FoodConsumption/FoodAvailIndex.htm

- U.S. Food and Drug Administration (nutrition labeling, antibiotics used in livestock, contaminants in food)   www.fda.gov

Vegetarian diets

- Earthsave   www.earthsave.org

- Meatless Monday   www.meatlessmonday.org

- Physicians Committee for Responsible Medicine   www.pcrm.org

- Vegetarian Resource Group   www.vrg.org/

- Vegsource   www.vegsource.com

Where to buy or eat food that is locally grown or produced humanely

- Community-supported agriculture (subscriptions for local produce)   www.localharvest.org/csa/

- Farmers' markets   www.localharvest.org <br> ww.ams.usda.gov/farmersmarkets/

- Humane Farm Animal Care   www.certifiedhumane.org

- Locally grown meats   www.eatwellguide.com

Notes

Preface: Greener Diets for a Healthier World (pp. vii-xiv)

1. G. Eshel and P. Martin, "Diet, energy, and global warming," *Earth Interactions* (2005) 10(9):1–17.

2. I. Hoffmann, "Ecological impact of a high-meat, low-meat and ovo-lacto vegetarian diet," presentation at the Fourth International Congress on Vegetarian Nutrition, Loma Linda, CA, Apr. 2002.

3. M. Pollan, *The Omnivore's Dilemma: A Natural History of Four Meals* (East Rutherford, NJ: Penguin Press, 2006).

4. L.R. Brown, "Running on empty," *Forum Appl Res Pub Pol* (2001) 16(1):6–8.

5. R. Naylor, H. Steinfeld, W. Falcon, et al., "Losing the links between livestock and land," *Science* (2005) Dec. 9:1621–22.

The Fatted Steer (pp. 3-13)

1. Omaha Steaks, www.omahasteaks.com; Morton's Steakhouse, www.mortons.com; and Ultimate Entree, www.ultimateentree.com/mm5/merchant.mvc?Screen=prime.

2. Shula's, www.certifiedangusbeef.com/shulas/shula1.html.

3. U.S. Department of Agriculture, Agricultural Research Service (USDA ARS), "Working towards a consistently tender steak," *Agr Res* (2005) 53(2):8–9; www.ars.usda.gov/is/AR/archive/feb05/steak0205.htm.

4. G. Fry, "The importance of intramuscular fat" (Rose Bud, AR: Bovine Consulting and Engineering), www.bovineengineering.com/impt_intra_musc_fat.html.

5. USDA, Agricultural Marketing Service (USDA AMS), "National summary of meats graded, fiscal year 2004," www.ams.usda.gov/lsg/mgc/Reports/MNFY04.pdf.

6. USDA AMS, "Comparison of certified beef programs" (2006), www.ams.usda.gov/lsg/certprog/industry.htm.

7. USDA, Food Safety and Inspection Service (USDA FSIS), "Inspection & grading – what are the differences?," safe food handling fact sheet, www.fsis.usda.gov/Fact_Sheets/Inspection_&_Grading/index.asp.

8. National Cattlemen's Beef Association, "Beef By Products Usage" (1996), www.beef.org/uDocs/Beef%20By%20Products%20Usage%201996.doc.

9. I.B. Mandell, J.G. Buchanan-Smith, and C.P. Campbell, "Effects of forage vs grain feeding on carcass characteristics, fatty acid composition, and beef quality in Limousin-cross steers when time on feed is controlled," *J Anim Sci* (1998) 76:2619–30.

10. S.P. Greiner, *Beef Cattle Breeds and Biological Types*, Publication No. 400–803 (Blacksburg, VA: Virginia Cooperative Extension, 2002), www.ext.vt.edu/pubs/beef/400–803/400–803.html.

11. T.E. Engle and J.W. Spears, "Effect of finishing system (feedlot or pasture), high-oil maize, and copper on conjugated linoleic acid and other fatty acids in muscle of finishing steer," *Anim Sci* (2004) 78:261–69.

12. Mandell, Buchanan-Smith, and Campbell, "Effects of forage."

13. Engle and Spears, "Effect of finishing system."

14. F.L. Laborde, I.B. Mandell, J.J. Tosh, et al., "Breed effects on growth performance, carcass characteristics, fatty acid composition, and palatability attributes in finishing steers," *J Anim Sci* (2001) 79:355–65.

15. S.K. Duckett, D.G. Wagner, L.D. Yates, et al., "Effects of time on feed on beef nutrient composition," *J Anim Sci* (1993) 71:2079–88.

16. A useful summary of studies, which vary widely in design, comparing grass-fed and grain-fed beef is presented in K. Clancy, *Greener Pastures: How Grass-Fed Beef and Milk Contribute to Healthy Eating* (Cambridge, MA: Union of Concerned Scientists, 2006), www.ucsusa.org.

17. Iowa Corn Fed, www.iowacornfed.com/.

18. P. Letheby, "Organic grass-fed beef: more than a niche?," *Grand Island Independent* Aug. 1, 2003, www.theindependent.com/stories/080103/opi_pete01.shtml.

19. C. Kummer, "Back to grass," *Atlantic Monthly* May 2003:138–42.

20. P. Brewer and C. Calkins, "Quality traits of grain- and grass-fed beef: a review," *2003 Nebraska Beef Cattle Report* (University of Nebraska Cooperative Extension), http://beef.unl.edu/beefreports/200327.shtml.

21. I.B. Mandell, E.A. Gullett, J.W. Wilton, et al., "Effects of breed and dietary energy content within breed on growth performance, carcass and chemical composition and beef quality in Hereford and Simmental steers," *Can J An Sci* (1998) 78:535–38.

22. T.R. Neely, C.L. Lorenzen, R.K. Miller, et al., "Beef customer satisfaction: role of cut, USDA quality grade, and city on in-home consumer ratings," *J Anim Sci* (1998) 76:1027–33.

23. K. Severson, "Give 'em a chance, steers will eat grass," *New York Times* June 1, 2005:F1.

24. Cattlemen's Beef Board, "Beef: it's what's for dinner," www.beefitswhatsfordinner.com/index.asp.

25. American Grass Fed Beef, www.americangrassfedbeef.com/grass-fed-beef-steak.asp.

26. M. Pariza, "Perspective on the safety and effectiveness of conjugated linoleic acid," *Am J Clin Nutr* (2004) 79(6 Suppl):1132s–6s.

27. J.M. Gaullier, J. Halse, K. Hoye, et al., "Supplementation with conjugated linoleic acid for 24 months is well tolerated by and reduces body fat mass in healthy, overweight humans," *J Nutr* (2005) 135:778–84.

28. Y. Wang and P.J. Jones, "Dietary conjugated linoleic acid and body composition," *Am J Clin Nutr* (2004) 79:1153S–8S; and S. Desroches, P.Y. Chouinard, I. Galibois, et al., "Lack of effect of dietary conjugated linoleic acids naturally incorporated into butter on the lipid profile and body composition of overweight and obese men," *Am J Clin Nutr* (2005) 82:309–19.

29. D.S. Kelley and K.L. Erickson, "Modulation of body composition and immune cell functions by conjugated linoleic acid in humans and animal models: benefits vs. risks," *Lipids* (2003) 38:377–86; and D. Mozaffarian, M.B. Katan, A. Ascherio, et al., "Trans fatty acids and cardiovascular disease," *N Engl J Med* (2006) 345:1601–13.

30. Institute of Medicine (IOM), *Dietary Reference Intakes for Energy, Carbohydrate, Fiber, Fat, Fatty Acids, Cholesterol, Protein, and Amino Acids* (*Macronutrients*) (Washington, DC: National Academies Press, 2002), pp. 837–38.

31. See, for example, IOM, *Dietary Reference Intakes*; C. Ip and J.A. Scimeca, "Conjugated linoleic acid and linoleic acid are distinctive modulators of mammary carcinogenesis," *Nutr Cancer* (1997) 27:131–35; and S. Vissonneau, A. Cesano, S.A. Tepper, et al., "Conjugated linoleic acid suppresses the growth of human breast adenocarcinoma cells in SCID mice," *Anticancer Res* (1997) 17:969–74.

32. T. Friend, "Fatty acid aids war on cancer," *USA Today* Apr. 5, 1989:1D.

33. IOM, *Dietary Reference Intakes*.

34. The fat in grass-fed beef contains about three to six times as much CLA as that in grain-fed beef. Grass-fed beef has about half as much fat as grain-fed beef. Engle and Spears, "Effect of finishing system"; F. Martz, M. Weiss, R. Kallenbach, et al., *Conjugated Linoleic Acid Content of Pasture Finished Beef and Implications for Human Diets* (Columbia, MO: University of Missouri, 2004), www.farmprofitability.org/research/beef/linoleic.htm; D.C. Rule, K.S. Broughton, S.M. Shellito, et al., "Comparison of muscle fatty acid profiles and cholesterol concentrations of bison, beef cattle, elk, and chicken," *J Anim Sci* (2002) 80:1202–11; and C.S. Poulson, T.R. Dhiman, A.L. Ure, et al., "Conjugated linoleic acid content of beef from cattle fed diets containing high grain, CLA, or raised on forages," *Livestock Prod Sci* (2004) 91:117–28.

35. Rule et al., "Comparison of muscle fatty acid profiles."

36. P.M. Kris-Etherton, W.S. Harris, and L.J. Appel, "Omega-3 fatty acids and cardiovascular disease: new recommendations from the American Heart Association," *Arterioscl Thromb Vasc Biol* (2003) 23:151–52; and C. Wang, M. Chung, A. Lichtenstein, et al., *Effects of Omega-3 Fatty Acids on Cardiovascular Disease: Summary, Evidence Report/Technology Assessment No. 94*, AHRQ Publication No. 04-E009-1 (Rockville, MD: Agency for Healthcare Research and Quality, 2004).

37. G. Gerster, "Can adults adequately convert alpha-linolenic acid (18:3n-3) to eicosapentaenoic acid (20:5n-3) and docosapentaenoic (22:6n-3)?," *Int J Vitam Nutr Res* (1998) 68(3):159–73; G.C. Burdge and S.A. Wootton, "Conversion of alpha-linolenic acid to eicosapentaenoic, docosapentaenoic and docosahexaenoic acids in young women," *Br J Nutr* (2002) 88:411–20; and J.T. Brenna, "Efficiency of conversion of alpha-linolenic acid to long chain n-3 fatty acids in man," *Curr Opin Clin Nutr Metab Care* (2002) 5:127–32.

38. P.M. Kris-Etherton, W.S. Harris, and L.J. Appel, "American Heart Association Scientific Statement: fish consumption, fish oil, omega-3 fatty acids, and cardiovascular disease," *Circulation* (2002) 106:2747–57. [Published correction appears in *Circulation* (2003) 107:512.]

39. An uncooked grass-fed rib steak contains about 13 milligrams of eicosapentanenoic acid (EPA) and 2 milligrams of docosahexaenoic acid (DHA) per 3½ ounces. It also contains about 33 milligrams of alpha-linolenic acid per serving, which provides the body with no more than 8 milligrams of EPA and DHA. Thus, a 7-ounce uncooked rib steak could provide, at most, about 46 milligrams of EPA and DHA. Certain other cuts have twice as much omega-3s. J.D. Wood, M. Enser, A.V. Fisher, et al., "Animal nutrition and metabolism group symposium on improving meat production for future needs," *Proc Nutr Soc* (1999) 58:363–70.

40. Cleveland Clinic, *The Power of Fish* (Cleveland, 2003), www.clevelandclinic.org/heartcenter/pub/guide/prevention/nutrition/omega3.htm.

41. USDA, Economic Research Service, "Briefing room: land use, value, and management: major uses of land" (2002), www.ers.usda.gov/Briefing/LandUse/majorlandusechapter.htm, accessed Dec. 27, 2005.

42. U.S. Environmental Protection Agency (EPA), *Inventory of U.S. Greenhouse Gas Emissions and Sinks: 1990–1998*, EPA 236-R-00-001 (2000), http://yosemite.epa.gov/oar/globalwarming.nsf/UniqueKeyLookup/SHSU5BMQ76/$File/2000-inventory.pdf, p. K-8.

43. American Society of Agricultural Engineers, *Manure Production and Characteristics* (St. Joseph, MI, 2002), pp. 687–89.

44. K. Richardson and P.A. McKay, "On the farm, chickens come home to roost," *Wall Street Journal* Aug. 12, 2005:C1.

Argument #1. Less Chronic Disease and Better Overall Health (pp. 17-57)

1. J.M. McGinnis and W.H. Foege, "The immediate vs the important," *JAMA* (2004) 291:1263–64. Their estimated range for 2000 was 340,000 to 642,000 deaths per year, or 16 to 30 percent of all deaths.

2. M.M. Miniño, E. Arias, K.D. Kochanek, et al., "Deaths: final data for 2000," *Natl Vital Stat Rep* (2002) 50(15):1–120.

3. *Morbidity and Mortality Weekly*, "Trends in intake of energy and macronutrients: United States, 1971–2000," *MMWR* (2004) 53:80–82; and J. Putnam, J. Allshouse, and L.S. Kantor, "U.S. per capita food supply trends: more calories, refined carbohydrates, and fats," *FoodReview* (2002) 25(3):2–15.

4. U.S. Department of Agriculture, Office of Communications (USDA OC), *Agriculture Fact Book 2001–2002* (2003), www.usda.gov/factbook/chapter2.htm.

5. USDA, National Agricultural Statistics Service (USDA NASS), *Milk Production, Disposition, and Income 2002 Summary* (Washington, DC, 2003), p. 2; USDA NASS, *Poultry Slaughter 2002 Annual Summary* (Washington, DC, 2003), p. 2; USDA NASS, *Livestock Slaughter 2002 Summary* (Washington, DC, 2003), pp. 35, 41, 49; USDA NASS, *Chickens and Eggs 2003 Summary* (Washington, DC, 2004), p. 2; and USDA, Economic Research Service (USDA ERS), Food Availability database, www.ers.usda.gov/Data/FoodConsumption/FoodAvailQueriable.aspx#midForm.

6. USDA ERS, *Food Availability (Per Capita)* (2005), www.ers.usda.gov/data/foodconsumption/FoodAvailIndex.htm.

7. USDA OC, *Agriculture Fact Book*.

8. U.S. Department of Health and Human Services and U.S. Department of Agriculture (DHHS/USDA), *Dietary Guidelines for Americans* (2005), www.health.gov/dietaryguidelines/dga2005/document/pdf/DGA2005.

9. P.A. Cotton, A.F. Subar, J.E. Friday, et al., "Dietary sources of nutrients among US adults, 1994 to 1996," *J Am Diet Assoc* (2004) 104:921–30. Food consumption data from

USDA Agricultural Research Service (USDA ARS), Food Surveys Research Group, Continuing Survey of Food Intakes by Individuals 1994–1996, www.barc.usda.gov/bhnrc/foodsurvey/home.htm.

10. A. Keys, J.T. Anderson, and F. Grande, "Serum cholesterol response to changes in the diet. IV. Particular saturated fatty acids in the diet," *Metabolism* (1965) 65:776–87; D.M. Hegsted, L.M. Ausman, J.A. Johnson, et al., "Dietary fat and serum lipids: an evaluation of the experimental data," *Am J Clin Nutr* (1993) 57:875–83; R.P. Mensink and M.B. Katan, "Effect of dietary fatty acids on serum lipids and lipoproteins: a meta-analysis of 27 trials," *Arterioscler Thromb* (1992) 12:911–19; and P.M. Kris-Etherton, A.E. Binkoski, G. Zhao, et al., "Dietary fat: assessing the evidence in support of a moderate-fat diet; the benchmark based on lipoprotein metabolism," *Proc Nutr Soc* (2002) 61(2):287–98.

11. J.E. Manson, H. Tosteson, P.M. Ridker, et al., "The primary prevention of myocardial infarction," *N Engl J Med* (1992) 326:1406–16.

12. M. Jacobson and H. D'Angelo, "Heart disease deaths caused by animal foods," unpublished report (Washington, DC: Center for Science in the Public Interest [CSPI], 2006). Estimates based on the four different research groups' formulas ranged from 30,000 to 107,000 deaths per year.

13. The $1 trillion sum is the present value discounted at 3 percent. It is based on the U.S. Food and Drug Administration's (FDA's) estimate of the health and economic benefits of lowering dietary levels of trans fat, which have adverse effects on blood cholesterol levels and cause heart disease. See FDA, "Nutrition labeling," *Fed Reg* (1999) 64:62746–825.

14. American Heart Association (AHA), *Heart Disease and Stroke Statistics: 2005 Update* (Dallas, 2005), p. 51; and L.S. Longwell, communications department, IMS Health, Inc., response to CSPI data request, Oct. 25, 2004.

15. Longwell, response.

16. American Cancer Society, *Cancer Facts and Figures, 2004* (Atlanta, 2004); AHA, *Heart Disease and Stroke*; American Diabetes Association, "Economic costs of diabetes in the U.S. in 2002," *Diabetes Care* (2003) 26:917–32; and E. Frazão, *America's Eating Habits: Changes and Consequences*, Agriculture Information Bulletin No. 750 (1999), www.ers.usda.gov/publications/aib750/aib750a.pdf.

17. N.D. Barnard, A. Nicholson, and J.L. Howard, "The medical costs attributable to meat consumption," *Prev Med* (1995) 24:646–55 (adjusted to 2005 dollars by CSPI).

18. A.M. Wolf and G.A. Colditz, "Social and economic effects of body weight in the United States," *Am J Clin Nutr* (1996) 63(suppl):466S–69S (adjusted to 2005 dollars by CSPI); and E.A. Finkelstein, I.C. Fiebelkorn, and G. Wang, "State-level estimates of annual medical expenditures attributable to obesity," *Obes Res* (2004) 12:18–24.

19. USDA ERS, *Data: Food Guide Pyramid Servings* (2005), www.ers.usda.gov/data/foodconsumption/FoodGuideIndex.htm#servings.

20. J.F. Guthrie, *Understanding Fruit and Vegetable Choices: Economic and Behavioral Influences*, Agriculture Information Bulletin 792-1, www.ers.usda.gov/publications/aib792/aib792-1/aib792-1.pdf.

21. USDA ERS, *Food Availability.*

22. G.E. Fraser, *Diet, Life Expectancy, and Chronic Disease: Studies of Seventh-day Adventists and Other Vegetarians* (New York: Oxford, 2003), p. 5; G.E. Fraser, "Associations between diet and cancer, ischemic heart disease, and all-cause mortality in non-Hispanic white California Seventh-day Adventists, *Am J Clin Nutr* (1999) 70(suppl):532S–38S; G.E. Fraser, P.W. Dysinger, C. Best, et al., "IHD risk factors in middle-aged Seventh-day Adventist men and their neighbors," *Am J Epidemiol* (1987) 126:638–46.

23. Fraser, Dysinger, Best, et al., "IHD risk factors."

24. Fraser, *Diet*, p. 13.

25. Fraser, "Associations."

26. Fraser, "Associations."

27. G.E. Fraser, J. Sabate, W.L. Beeson, et al., "A possible protective effect of nut consumption on risk of coronary heart disease: the Adventist Health Study," *Arch Intern Med* (1992) 152:1416–24.

28. Fraser, "Associations."

29. J. Berkel and F. de Waard, "Mortality pattern and life expectancy of Seventh-day Adventists in the Netherlands," *Int J Epidemiol* (1983) 12(4):455–59, cited in Fraser, *Diet*, p. 23.

30. M.L. Toohey, M.A. Haris, D. Williams, et al., "Cardiovascular disease risk factors are lower in African-American vegans compared to lacto-ovo vegetarians," *J Am Coll Nutr* (1998) 17:425–34.

31. Fraser, "Associations."

32. Fraser, *Diet*, pp. 141–42.

33. N. Brathwaite, H.S. Fraser, N. Modeste, et al., "Obesity, diabetes, hypertension, and vegetarian status among Seventh-day Adventists in Barbados: preliminary results," *Ethn Dis* (2003) 13:34–9; and Fraser, "Associations."

34. Fraser, "Associations."

35. Fraser, "Associations."

36. E.H. Haddad and J.S. Tanzman, "What do vegetarians in the United States eat?," *Am J Clin Nutr* (2003) 78(suppl):626S–32S.

37. E.T. Kennedy, S.A. Bowman, I.T. Spence, et al., "Popular diets: correlation to health, nutrition, and obesity," *J Am Diet Assoc* (2001) 101:411–20.

38. G.K. Davey, E.A. Spencer, P.N. Appleby, et al., "EPIC-Oxford lifestyle characteristics and nutrient intakes in a cohort of 33,993 meat-eaters and 31,546 non-meat-eaters in the UK," *Public Health Nutr* (2003) 6:259–69.

39. P.N. Appleby, M. Thorogood, J.I. Mann, et al., "The Oxford Vegetarian Study: an overview," *Am J Clin Nutr* (1999) 70(suppl):525S–31S.

40. Appleby et al., "Oxford Vegetarian Study."

41. Appleby et al., "Oxford Vegetarian Study."

42. Appleby et al., "Oxford Vegetarian Study"; and Fraser, *Diet*, pp. 233–35.

43. P.N. Appleby, G.K. Davey, and T.J. Key, "Hypertension and blood pressure among meat eaters, fish eaters, vegetarians and vegans in EPIC-Oxford," *Public Health Nutr* (2002) 5:645–54.

44. Appleby, Davey, and Key, "Hypertension."

45. Fraser, *Diet*, p. 220.

46. T. Key and G. Davey, "Prevalence of obesity is low in people who do not eat meat," *BMJ* (1996) 313:816–17.

47. P.N. Appleby, M. Thorogood, and J.I. Mann, "Low body mass index in non-meat eaters: the possible roles of animal fat, dietary fibre and alcohol," *Int J Obesity* (1998) 22(5):454–60.

48. P.K. Newby, K.L. Tucker, and A. Wolk, "Risk of overweight and obesity among semivegetarian, lactovegetarian, and vegan women," *Am J Clin Nutr* (2005) 81:1267–74.

49. T.J. Key, G.E. Fraser, M. Thorogood, et al., "Mortality in vegetarians and non-vegetarians: detailed findings from a collaborative analysis of 5 prospective studies," *Am J Clin Nutr* (1999) 70(suppl):516S–24S.

50. T.J. Key, G.K. Davey, and P.N. Appleby, "Health benefits of a vegetarian diet," *Proc Nutr Soc* (1999) 58(2):271–75.

51. Appleby, Thorogood, and Mann, "Low body mass index"; and Key, Davey, and Appleby, "Health benefits."

52. Fraser, *Diet*, pp. 236–38.

53. T.T. Fung, W.C. Willett, M.J. Stampfer, et al., "Dietary patterns and the risk of coronary heart disease in women," *Arch Intern Med* (2001) 161:1857–62.

54. F.B. Hu, E.B. Rimm, M.J. Stampfer, et al., "Prospective study of major dietary patterns and risk of coronary heart disease in men," *Am J Clin Nutr* (2000) 72:912–21.

55. I.L. Rouse, L.J. Beilin, D.P. Mahoney, et al., "Nutrient intake, blood pressure, serum and urinary prostaglandins and serum thromboxane B2 in a controlled trial with a lacto-ovo-vegetarian diet," *J Hypertens* (1986) 4:241–50; and S.E. Sciarrone, M.T. Strahan, L.J. Beilin, et al., "Biochemical and neurohormonal responses to the introduction of a lacto-ovo vegetarian diet," *J Hypertens* (1993) 11:849–60.

56. Rouse et al., "Nutrient intake"; and L.J. Appel, T.J. Moore, E. Obarzanek, et al., "A clinical trial of the effects of dietary patterns on blood pressure," *N Engl J Med* (1997) 336:1117–24.

57. A.M. Lees, A.Y. Mok, R.S. Lees, et al., "Plant sterols as cholesterol-lowering agents: clinical trials in patients with hypercholesterolemia and studies of sterol balance," *Atherosclerosis* (1977) 28:325–38; and D.J. Jenkins, T.M. Wolever, A.V. Rao, et al., "Effect on blood lipids of very high intakes of fiber in diets low in saturated fat and cholesterol," *N Engl J Med* (1993) 329:21–6.

58. D.J. Jenkins, C.W. Kendall, A. Marchie, et al., "Effects of a dietary portfolio of cholesterol-lowering foods vs lovastatin on serum lipids and C-reactive protein," *JAMA* (2003) 290:502–10; D.J. Jenkins, C.W. Kendall, A. Marchie, et al., "Direct comparison of a dietary portfolio of cholesterol-lowering foods with a statin in hypercholesterolemic participants," *Am J Clin Nutr* (2005) 81(2):380–87; and D.J. Jenkins, C.W. Kendall, A. Marchie, et al., "The effect of combining plant sterols, soy protein, viscous fibers, and almonds in treating hypercholesterolemia," *Metabolism* (2003) 52(11):1478–83.

59. S.M. Grundy, J.I. Cleeman, B.C.N. Merz, et al., "Implications of recent clinical trials for the National Cholesterol Education Program Adult Treatment Panel III Guidelines," *Circulation* (2004) 110:227–39.

60. H.A. Diehl, "Coronary risk reduction through intensive community-based lifestyle intervention: the Coronary Health Improvement Project (CHIP) experience," *Am J Cardiol* (1998) 82(10B).83T–87T.

61. D.W. Harsha, P.-H. Lin, E. Obarzanek, et al., "Dietary Approaches to Stop Hypertension: a summary of results," *J Am Diet Assoc* (1999) 99(suppl):S53–59; N.M. Karanja, E. Obarzanek, P.-H. Lin, et al., "Descriptive characteristics of the dietary patterns used in the Dietary Approaches to Stop Hypertension trial," *J Am Diet Assoc* (1999) 99(suppl): S19–S27; F.M. Sacks, L.P. Svetkey, W.M. Vollmer, et al., "Effects on blood pressure of reduced dietary sodium and the Dietary Approaches to Stop Hypertension (DASH) diet," *N Engl J Med* (2001) 344:3–10; Appel et al., "A clinical trial"; and E. Obarzanek, F.M. Sacks, and W.M. Vollmer, "Effects on blood lipids of a blood pressure-lowering diet: the Dietary Approaches to Stop Hypertension (DASH) trial," *Am J Clin Nutr* (2001) 74(1):80–89.

62. M. deLorgeril, P. Salen, J.-L. Martin, et al., "Mediterranean diet, traditional risk factors, and the rate of cardiovascular complications after myocardial infarction: final report of the Lyon Diet Heart Study," *Circulation* (1999) 99:779–85.

63. A. Leaf, "Dietary prevention of coronary heart disease: the Lyon Diet Heart Study," *Circulation* (1999) 99:733–35.

64. S.G. Aldana, R.L. Greenlaw, H.A. Diehl, et al., "Effects of an intensive diet and physical activity modification program on the health risks of adults," *J Am Diet Assoc* (2005) 105(3):371–81.

65. D. Ornish, S.E. Brown, L.W. Schenwitz, et al., "Can lifestyle changes reverse coronary heart disease?," *Lancet* (1990) 336:129–33; and D. Ornish, L.W. Schenwitz, J.H. Billings, et al., "Intensive lifestyle changes for reversal of coronary heart disease," *JAMA* (1998) 280:2001–07.

66. J. Koertge, G. Weidner, M. Elliott-Eller, et al., "Improvement in medical risk factors and quality of life in women and men with coronary artery disease in the Multicenter Lifestyle Demonstration Project," *Am J Cardiol* (2003) 91:1316–22.

67. D. Ornish, G. Weidner, W.R. Fair, et al., "Intensive lifestyle changes may affect the progression of prostate cancer," *J Urol* (2005) 174:1065–69.

68. C.B. Esselstyn Jr., "Updating a 12-year experience with arrest and reversal therapy for coronary heart disease (an overdue requiem for palliative cardiology)," *Am J Cardiol* (1999) 84(3):339–41; C.B. Esselstyn Jr., "Resolving the coronary artery disease epidemic through plant-based nutrition," *Prev Cardiol* (2001) 4:171–77; and "Becoming heart attack proof" (VegSource Interactive, Inc., 2003), www.vegsource.com/esselstyn/index.htm.

69. Institute of Medicine (IOM), *Dietary Reference Intakes for Energy, Carbohydrate, Fiber, Fat, Fatty Acids, Cholesterol, Protein, and Amino Acids* (*Macronutrients*) (Washington, DC: National Academies Press, 2002), pp. 297–302, 777–87; and Ornish et al., "Intensive lifestyle changes for reversal."

70. R.J. Barnard, T. Jung, and S.B. Inkeles, "Diet and exercise in the treatment of NIDDM: the need for early emphasis," *Diabetes Care* (1994) 17:1469–72.

71. M.G. Crane and C. Sample, "Regression of diabetic neuropathy with total vegetarian (vegan) diet," *J Nutr Med* (1994) 4:431–39.

72. American Cancer Society (ACS), "The complete guide – nutrition and physical activity," www.cancer.org/docroot/PED/content/PED_3_2X_Diet_and_Activity_Factors_That_Affect_Risks.asp?sitearea=PED; DHHS/USDA, *Dietary Guidelines for Americans; and* World Health Organization/Food and Agriculture Organization (WHO/FAO), *Diet, Nutrition and the Prevention of Chronic Diseases,* WHO Technical Report Series 916 (Geneva, 2003).

73. H.C. Hung, K.J. Joshipura, R. Jiang, et al., "Fruit and vegetable intake and risk of major chronic disease," *J Natl Cancer Inst* (2004) 96:1577–84; S. Liu, I.M. Lee, U. Ajani, et al., "Intake of vegetables rich in carotenoids and risk of coronary heart disease in men: the Physicians' Health Study," *Int J Epidemiol* (2001) 30:130–35; and L.A. Bazzano, J. He, L.G. Ogden, et al., "Fruit and vegetable intake and risk of cardiovascular disease in US adults: the first National Health and Nutrition Examination Survey Epidemiologic Follow-up Study," *Am J Clin Nutr* (2002) 76:93–99.

74. K.J. Joshipura, F.B. Hu, J.E. Manson, et al., "The effect of fruit and vegetable intake on risk for coronary heart disease," *Ann Intern Med* (2001) 134:1106–14; and S. Liu, J.E. Manson, I.M. Lee, et al., "Fruit and vegetable intake and risk of cardiovascular disease: the Women's Health Study," *Am J Clin Nutr* (2000) 72:922–28.

75. T.H. Rissanen, S. Voutilainen, J.K. Virtanen, et al., "Low intake of fruits, berries and vegetables is associated with excess mortality in men: the Kuopio Ischaemic Heart Disease Risk Factor (KIHD) Study," *J Nutr* (2003) 133:199–204.

76. Bazzano et al., "Fruit and vegetable intake."

77. P. Van't Veer, M.C. Jansen, M. Klerk, et al., "Fruits and vegetables in the prevention of cancer and cardiovascular disease," *Public Health Nutr* (2000) 3:103–07.

78. L.M. Steffen, C.H. Kroenke, X. Yu, et al., "Associations of plant food dairy product, and meat intakes with 15-y incidence of elevated blood pressure in young black and white adults: the Coronary Artery Risk Development in Young Adults (CARDIA) Study," *Am J Clin Nutr* (2005) 82:1169–77; and L. Dauchet, P. Amouyel, and J. Dallongeville, "Fruit and vegetable consumption and risk of stroke: a meta-analysis of cohort studies," *Neurology* (2005) 65:1193–97.

79. E. Riboli and T. Norat, "Epidemiologic evidence of the protective effect of fruit and vegetables on cancer risk," *Am J Clin Nutr* (2003) 78(suppl):559S–69S; T.J. Key, A. Schatzkin, W.C. Willett, et al., "Diet, nutrition, and the prevention of cancer," *Public Health Nutr* (2004) 7:187–200; and WHO/FAO, *Diet, Nutrition*.

80. Key et al., "Diet, nutrition"; WHO/FAO, *Diet, Nutrition*; C.H. van Gils, P.H. Peeters, H.B. Bueno-de-Mesquita, et al., "Consumption of vegetables and fruits and risk of breast cancer," *JAMA* (2005) 293:183–93; D. Feskanich, R.G. Ziegler, D.S. Michaud, et al., "Prospective study of fruit and vegetable consumption and risk of lung cancer among men and women," *J Natl Cancer Inst* (2000) 92:1812–23; L.E. Voorrips, R.A. Goldbohm, D.T. Verhoeven, et al., "Vegetable and fruit consumption and lung cancer risk in the Netherlands Cohort Study on diet and cancer," *Cancer Causes Control* (2001) 11:101–15; S.A. Smith-Warner, D. Spiegelman, S.S. Yaun, et al., "Intake of fruits and vegetables and risk of breast cancer: a pooled analysis of cohort studies," *JAMA* (2001) 285:769–76; M.P. Zeegers, R.A. Goldbohm, and P.A. van den Brandt, "Consumption of vegetables and fruits and urothelial cancer incidence: a prospective study," *Cancer Epid Biomarkers Prev* (2001) 10:1121–28; and D.S. Michaud, D. Spiegelman, S.K. Clinton, et al., "Fruit and vegetable intake and incidence of bladder cancer in a male prospective cohort," *J Natl Cancer Inst* (1999) 91:605–13.

81. Key et al., "Diet, nutrition"; and WHO/FAO, *Diet, Nutrition*.

82. ACS, "Complete Guide"; DHHS/USDA, *Dietary Guidelines for Americans*; and WHO/FAO, *Diet, Nutrition*.

83. W.C. Willett, "Harvesting the fruits of research: new guidelines on nutrition and physical activity," *CA Cancer J Clin* (2002) 52:66–67.

84. B.J. Rolls, J.A. Ello-Martin, and B.C. Tohill, "What can intervention studies tell us about the relationship between fruits and vegetable consumption and weight management?," *Nutr Rev* (2004) 62:1–17.

85. B.C. Tohill, "Fruit and vegetables and weight management," www.hkresources.org/articles/fruit_vegetable.ppt.

86. B.C. Tohill, J. Seymour, M. Serdula, et al., "What epidemiologic studies tell us about the relationship between fruit and vegetable consumption and body weight," *Nutr Rev* (2004) 62:365–74.

87. M.K. Serdula, C. Gillespie, L. Kettel-Khan, et al., "Trends in fruit and vegetable consumption among adults in the United States: behavioral risk factor surveillance system, 1994–2000," *Am J Pub Health* (2004) 94:1014–18.

88. Rolls, Ello-Martin, and Tohill, "What can intervention studies tell."

89. E.S. Ford and A.H. Mokdad, "Fruit and vegetable consumption and diabetes mellitus incidence among U.S. adults," *Prev Med* (2001) 32:33–9; K.L. Tucker, H. Chen, M.T. Hannan, et al., "Bone mineral density and dietary patterns in older adults: the Framingham Osteoporosis Study," *Am J Clin Nutr* (2002) 76:245–52; S.A. New, "Intake of fruits and vegetables: implications for bone health," *Proc Nutr Soc* (2003) 62:889–99; S.A. New, S.P. Robins, M.K. Campbell, et al., "Dietary influences on bone mass and bone metabolism: further evidence of a positive link between fruit and vegetable consumption and bone health?," *Am J Clin Nutr* (2000) 71:142–51; and K.L. Tucker, M.T. Hannan, H. Chen, et al., "Potassium and fruit and vegetables are associated with greater bone mineral density in elderly men and women," *Am J Clin Nutr* (1999) 69:727–36.

90. J.K. Campbell, K. Canene-Adams, B.L. Lindshield, et al., "Tomato phytochemicals and prostate cancer risk," *J Nutr* (2004) 134(12 suppl):3486S–92S.

91. S. Mannisto, S.A. Smith-Warner, D. Spiegelman, et al., "Dietary carotenoids and risk of lung cancer in a pooled analysis of seven cohort studies," *Cancer Epidemiol Biomarkers Prev* (2004) 13:40–48.

92. Michaud et al., "Fruit and vegetable intake."

93. C.S. Johnston, C.A. Taylor, and J.S. Hampl, "More Americans are eating "5 A Day" but intakes of dark green and cruciferous vegetables remain low," *J Nutr* (2000) 130:3063–67; and DHHS/USDA, *Dietary Guidelines for Americans*.

94. WHO/FAO, *Diet, Nutrition*.

95. 2005 Dietary Guidelines Advisory Committee, *2005 Dietary Guidelines Advisory Committee Report*, www.health.gov/dietaryguidelines/dga2005/report/, part D, sec. 6.

96. DHHS/USDA, *Dietary Guidelines for Americans*.

97. USDA OC, *Agriculture Fact Book*.

98. Fraser et al., "A possible protective effect"; D.R. Jacobs Jr., K.A. Meyer, L.H. Kushi, et al., "Whole-grain intake may reduce the risk of ischemic heart disease death in post-menopausal women: the Iowa Women's Health Study," *Am J Clin Nutr* (1998) 68:248–57; S. Liu, M.J. Stampfer, F.B. Hu, et al., "Whole-grain consumption and risk of coronary heart disease: results from the Nurses' Health Study," *Am J Clin Nutr* (1999) 70:412–19; S. Liu, J.E. Manson, M.J. Stampfer, et al., "Whole grain consumption and risk of ischemic stroke in women: a prospective study," *JAMA* (2000) 284:1534–40; and J.W. Anderson, "Whole grains protect against atherosclerotic cardiovascular disease," *Proc Nutr Soc* (2003) 62:135–42.

99. M.A. Murtaugh, D.R. Jacobs Jr., B. Jacob, et al., "Epidemiological support for the protection of whole grains against diabetes," *Proc Nutr Soc* (2003) 62:143–49.

100. M.A. Pereira, D.R. Jacobs Jr., J.J. Pins, et al., "Effect of whole grains on insulin sensitivity in overweight hyperinsulinemic adults," *Am J Clin Nutr* (2002) 75:848–55.

101. IOM, *Dietary Reference Intakes (Macronutrients)*, p. 370.

102. W.H. Aldoori, E.L. Giovannucci, H.R. Rockett, et al., "A prospective study of dietary fiber types and symptomatic diverticular disease in men," *J Nutr* (1998) 128(4):714–19; and IOM, *Dietary Reference Intakes (Macronutrients)*, p. 372.

103. Fraser et al., "A possible protective effect"; J. Sabate, G.E. Fraser, K. Burke, et al., "Effects of walnuts on serum lipid levels and blood pressure in normal men," *N Engl J Med* (1993) 328: 603–07; L.H. Kushi, A.R. Folsom, R.J. Prineas, et al., "Dietary antioxidant vitamins and death from coronary heart disease in postmenopausal women," *N Engl J Med* (1996) 334:1156–62; F.B. Hu, M.J. Stampfer, J.E. Manson, et al., "Frequent nut consumption and risk of coronary heart disease in women: prospective cohort study," *BMJ* (1998) 317:1341–45; J. Sabate, "Nut consumption, vegetarian diets, ischemic heart disease risk, and all-cause mortality: evidence from epidemiologic studies," *Am J Clin Nutr* (1993) 70(suppl):500S–03S; G.A. Spiller, D.A. Jenkins, O. Bosello, et al., "Nuts and plasma lipids: an almond-based diet lowers LDL-C while preserving HDL-C," *J Am Coll Nutr* (1998) 17(3):285–90; and Jenkins et al., "Effects of a dietary portfolio."

104. J. Mukuddem-Petersen, W. Oosthuizen, and J.C. Jerling, "A systematic review of the effects of nuts on blood lipid profiles in humans," *J Nutr* (2005) 135:2082–89; P.M. Kris-Etherton, S. Yu-Poth, J. Sabate, et al., "Nuts and their bioactive constituents: effects on serum lipids and other factors that affect disease risk," *Am J Clin Nutr* (1999) 70(suppl):504S–11S; Fraser, *Diet*, p. 257; and FDA, "Qualified health claims: letter of enforcement discretion—nuts and coronary heart disease," Docket No. 02P-0505 (2003), www.cfsan.fda.gov/~dms/qhcnuts2.html.

105. USDA ERS, *Food Availability*.

106. L.A. Bazzano, J. He, L.G. Ogden, et al., "Legume consumption and risk of coronary heart disease in US men and women," *Arch Intern Med* (2001) 161:2573–78.

107. J.W. Anderson and A.W. Major, "Pulses and lipaemia, short- and long-term effect potential in the prevention of cardiovascular disease," *Br J Nutr* (2002) 88(suppl):S263–71.

108. F.M. Sacks, A. Lichtenstein, L. Van Horn, et al., "Soy protein, isoflavones, and cardiovascular health: an American Heart Association science advisory for professionals from the Nutrition Committee," *Circulation* (2006) 113:1034–44.

109. P.H. Peeters, L. Boker Keinan, Y.T. van der Schouw, et al., "Phytoestrogens and breast cancer risk: review of the epidemiological evidence," *Breast Cancer Res Trea* (2003) 77:171–83; M.J. Messina, "Emerging evidence on the role of soy in reducing prostate cancer risk," *Nutr Rev* (2003) 61:117–31; and Key et al., "Diet, nutrition."

110. S. Kreijkamp-Kaspers, L. Kok, D.E. Grobbee, et al., "Effect of soy protein containing isoflavones on cognitive function, bone mineral density, and plasma lipids in postmenopausal women," *JAMA* (2004) 292:65–74.

111. Bazzano et al., "Legume consumption."

112. I. Darmadi-Blackberry, M.L. Wahlqvist, A. Kouris-Blazos, et al., "Legumes: the most important dietary predictor of survival in older people of different ethnicities," *Asia Pac J Clin Nutr* (2004) 6:217–20.

113. Cotton et al., "Dietary sources of nutrients." Note that "ice cream" also includes sherbet and frozen yogurt, and "cakes, cookies" also includes quick breads and donuts.

114. IOM, *Dietary Reference Intakes (Macronutrients)*, p. 422.

115. These studies link saturated fat to diabetes: F.B. Hu, R.M. van Dam, and S. Liu, "Diet and risk of type II diabetes: the role of types of fat and carbohydrate," *Diabetologia* (2001) 44:805–17; E.J.M. Feskens, S.M. Virtanen, L. Rasanen, et al., "Dietary factors determining diabetes and impaired glucose tolerance: a 20-year follow-up of the Finnish and Dutch cohorts of the Seven Countries Study," *Diabetes Care* (1995) 18:1104–12; and D.R. Parker, S.T. Weiss, R. Troisi, et al., "Relationship of dietary saturated fatty acids and body habitus to serum insulin concentrations: the Normative Aging Study," *Am J Clin Nutr* (1993) 58:129–36.

Other studies find no such links: M.B. Costa, S.R.G. Ferreira, L.J. Franco, et al., "Dietary patterns in a high-risk population for glucose intolerance," *J Epidemiol* (2000) 10:111–17; and J. Salmeron, F.B. Hu, E. Manson, et al., "Dietary fat intake and risk of type 2 diabetes in women," *Am J Clin Nutr* (2001) 73:1019–26.

116. IOM, *Dietary Reference Intakes (Macronutrients)*, pp. 422–23.

117. DHHS/USDA, *Dietary Guidelines for Americans.*

118. IOM, *Dietary Reference Intakes (Macronutrients)*, p. 542; M. Tanasescu, E. Cho, J.E. Manson, and F.B. Hu, "Dietary fat and cholesterol and the risk of cardiovascular disease," *Am J Clin Nutr* (2004) 79:999–1005; and A. Ascherio, E.B. Rimm, E.L. Giovannucci, et al., "Dietary fat and risk of coronary heart disease in men: cohort follow up study in the United States," *BMJ* (1996) 313:84–90.

119. IOM, *Dietary Reference Intakes (Macronutrients)*, p. 1058.

120. DHHS/USDA, *Dietary Guidelines for Americans.*

121. F.M. Sacks, A. Donner, W.P. Castelli, et al., "Effect of ingestion of meat on plasma cholesterol of vegetarians," *JAMA* (1981) 246:640–44.

122. Fraser, "Associations."

123. Sacks et al., "Effect of ingestion"; A. Ascherio, C. Hennekens, W.C. Willett, et al., "Prospective study of nutritional factors, blood pressure, and hypertension among US women," *Hypertension* (1996) 27:1065–72; and Steffen et al., "Associations of plant food."

124. ACS, "Complete guide"; and WHO/FAO, *Diet, Nutrition.*

125. A. Chao, M.J. Thun, C.J. Connell, et al., "Meat consumption and risk of colorectal cancer," *JAMA* (2005) 293:172–82.

126. Fraser, "Associations."

127. T. Norat, S. Bingham, P. Ferrari, et al., "Meat, fish, and colorectal cancer risk: the European Prospective Investigation into Cancer and Nutrition," *J Natl Cancer Inst* (2005) 97:906–16.

128. T. Norat, A. Lukanova, P. Ferrari, et al., "Meat consumption and colorectal cancer risk: dose-response meta-analysis of epidemiological studies," *Int J Cancer* (2002) 98:241–56.

129. M.S. Sandhu, I.R. White, and K. McPherson, "Systematic review of the prospective cohort studies on meat consumption and colorectal cancer risk: a meta-analytical approach," *Cancer Epidemiol Biomarkers Prev* (2001) 10:439–46.

130. U. Nöthlings, L.R. Wilkens, S.P. Murphy, et al., "Meat and fat intake as risk factors for pancreatic cancer: the multiethnic cohort study," *J Natl Cancer Inst* (2005) 97(19):1458–65.

131. R.M. van Dam, W.C. Willett, E.B. Rimm, et al., "Dietary fat and meat intake in relation to risk of type 2 diabetes in men," *Diabetes Care* (2002) 25:417–24; T.T. Fung, M. Schulze, J.E. Manson, et al., "Dietary patterns, meat intake, and the risk of type 2 diabetes in women," *Arch Intern Med* (2004) 164:2235–40.

132. DHHS/USDA, *Dietary Guidelines for Americans*; R.P. Heaney, "Calcium, dairy products and osteoporosis," *J Am Coll Nutr* (2000) 19:835S–9S; and R.L. Weinsier and C.L. Krumdieck, "Dairy foods and bone health: examination of the evidence," *Am J Clin Nutr* (2000) 72:681–89.

133. S. Mizuno, K. Matsuura, T. Gotou, et al., "Antihypertensive effect of casein hydrolysate in a placebo-controlled study in subjects with high-normal blood pressure and mild hypertension," *Br J Nutr* (2005) 94:84–91.

134. Weinsier and Krumdieck, "Dairy foods."

135. 2005 Dietary Guidelines Advisory Committee, *Report*, part D, sec. 6; Appel et al., "A clinical trial"; and L.J. Massey, "Dairy food consumption, blood pressure and stroke," *J Nutr* (2001) 131:1875–78.

136. Cotton et al., "Dietary sources of nutrients."

137. F.B. Hu, M.J. Stampfer, J.E. Manson, et al., "Dietary saturated fats and their food sources in relation to the risk of coronary heart disease in women," *Am J Clin Nutr* (1999) 70:1001–08.

138. X. Gao, M.P. LaValley, and K.L. Tucker, "Prospective studies of dairy product and calcium intakes and prostate cancer risk: a meta-analysis," *J Natl Cancer Inst* (2005) 97:1768–77; J.M. Chan and E.L. Giovannucci, "Dairy products, calcium, and vitamin D and risk of prostate cancer," *Epidemiol Rev* (2001) 23:87–92; and G. Severi, D.R. English, J.L. Hopper, et al., Letter to the editor re. "Prospective studies of dairy product and calcium intakes and prostate cancer risk: a meta-analysis," *J Nat Cancer Inst* (2006) 98:794–95.

139. Gao, LaValley, and Tucker, "Dairy product."

140. USDA ARS, *Pyramid Servings Intakes in the United States 1999–2002, 1 Day* (2005), http://usna.usda.gov/cnrg/services/ts_3-0.pdf.

141. M. Tseng, R.A. Breslow, B.I. Graubard, et al., "Dairy, calcium, and vitamin D intakes and prostate cancer risk in the National Health and Nutrition Examination Epidemiologic Follow-up Study cohort," *Am J Clin Nutr* (2005) 81:1147–54; and J.M. Chan, M.J. Stampfer, J. Ma, et al., "Dairy products, calcium, and prostate cancer risk in the Physicians' Health Study," *Am J Clin Nutr* (2001) 74:549–54.

142. These studies link calcium to prostate cancer: C. Rodriguez, M.L. McCullough, A.M. Mondul, et al., "Calcium, dairy products, and risk of prostate cancer in a prospective cohort of United States men," *Cancer Epidemiol Biomarkers Prev* (2003) 12:597–603; and Chan et al., "Dairy products." These studies dispute that claim: A.G. Schuurman, P.A. van den Brandt, E. Dorant, et al., "Animal products, calcium and protein and prostate cancer risk in the Netherlands Cohort study," *Br J Cancer* (1999) 80:1107–13; and J.M. Chan, P. Pietinen, M. Virtanen, et al., "Diet and prostate cancer risk in a cohort of smokers, with a specific focus on calcium and phosphorous," *Cancer Causes Control* (2000) 11:859–67.

143. T. Norat and E. Riboli, "Dairy products and colorectal cancer: a review of possible mechanisms and epidemiological evidence," *Eur J Clin Nutr* (2003) 57:1–17; and E. Cho, S.A. Smith-Warner, D. Spiegelman, et al., "Dairy foods, calcium, and colorectal cancer: a pooled analysis of 10 cohort studies," *J Natl Cancer Inst* (2004) 96:1015–22.

144. M.-H. Shin, M.D. Holmes, S.E. Hankinson, et al., "Intake of dairy products, calcium, and vitamin D and risk of breast cancer," *J Natl Cancer Inst* (2002) 94:1301–11; M.L. McCullough, C. Rodriguez, W.R. Diver, et al., "Dairy, calcium, and vitamin D intake and postmenopausal breast cancer risk in the Cancer Prevention Study II nutrition cohort," *Cancer Epidemiol Biomarkers Prev* (2005) 14(12):2898–904; and P.G. Moorman and P.D. Terry, "Consumption of dairy products and the risk of breast cancer: a review of the literature," *Am J Clin Nutr* (2004) 80:5–14.

145. Cotton et al., "Dietary sources of nutrients."

146. 2005 Dietary Guidelines Advisory Committee, *Report*, part D, sec. 4.

147. IOM, *Dietary Reference Intakes (Macronutrients)*, p. 563.

148. Appleby et al., "Oxford Vegetarian Study."

149. R.M. Weggemans, P.L. Zock, and M.B. Katan, "Dietary cholesterol from eggs increases the ratio of total cholesterol to high-density lipoprotein cholesterol in humans: a meta-analysis," *Am J Clin Nutr* (2001) 73:885–91.

150. F.B. Hu, M.J. Stampfer, E.B. Rimm, et al., "A prospective study of egg consumption and risk of cardiovascular disease in men and women," *JAMA* (1999) 281:1387–94.

151. K. He, Y. Song, M.L. Daviglus, et al., "Accumulated evidence on fish consumption and coronary heart disease mortality: a meta-analysis of cohort studies," *Circulation* (2004) 109:2705–11.

152. P.M. Kris-Etherton, W.S. Harris, and L.J. Appel, "Omega-3 fatty acids and cardiovascular disease: new recommendations from the American Heart Association," *Arteriscl Thromb Vasc Biol* (2003) 23:151–52; and IOM, *Dietary Reference Intakes (Macronutrients)*, pp. 828–29.

153. WHO/FAO, *Diet, Nutrition*; P.M. Kris-Etherton, W.S. Harris, and L.J. Appel, "American Heart Association Scientific Statement: fish consumption, fish oil, omega-3 fatty acids, and cardiovascular disease," *Circulation* (2002) 106:2747–57; DHHS/USDA, *Dietary Guidelines for Americans*; and 2005 Dietary Guidelines Advisory Committee, *Report*.

154. Norat et al., "Meat, fish, and colorectal cancer risk."

155. M.F. Leitzmann, M.J. Stampfer, D.C. Michaud, et al., "Dietary intake of n-3 and n-6 fatty acids and the risk of prostate cancer," *Am J Clin Nutr* (2004) 80:204–16; K. Augustsson, D.S. Michaud, E.B. Rimm, et al., "A prospective study of intake of fish and marine fatty acids and prostate cancer," *Cancer Epidemiol Biomarkers Prev* (2003) 12:64–7; P. Terry, P. Lichtenstein, M. Feychting, et al., "Fatty fish consumption and risk of prostate cancer," *Lancet* (2001) 357:1764–6; and A.E. Norrish, C.M. Skeaff, G.L. Arribas, et al., "Prostate cancer risk and consumption of fish oils: a dietary biomarker-based case-control study," *Br J Cancer* (1999) 81:1238–42.

156. IOM, *Dietary Reference Intakes (Macronutrients)*, p. 348.

157. IOM, *Dietary Reference Intakes (Macronutrients)*, pp. 351–61; and American Dietetic Association, "Health implications of dietary fiber," *J Am Diet Assoc* (2002) 102:993–1000.

158. IOM, *Dietary Reference Intakes (Macronutrients)*, pp. 370–72; and Aldoori et al., "Dietary fiber types."

159. B. Trock, E. Lanza, and P. Greenwald, "Dietary fiber, vegetables, and colon cancer: critical review and meta-analyses of the epidemiologic evidence," *J Natl Cancer Inst* (1990) 82:650–61.

160. Epidemiology study: Y. Park, D.J. Hunter, D. Spiegelman, et al., "Dietary fiber intake and risk of colon cancer: a pooled analysis of prospective cohort studies," *JAMA* (2005)

294:2849–57. Intervention studies: D.S. Alberts, M.E. Martinez, D.J. Roe, et al.,"Lack of effect of a high-fiber cereal supplement on the recurrence of colorectal adenomas," *N Engl J Med* (2000) 342:1156–62; C. Bonithon-Kopp, O. Kronborg, A. Giacosa, et al., "Calcium and fibre supplementation in prevention of colorectal adenoma recurrence: a randomized intervention trial," *Lancet* (2000) 356:1300–06; and A. Schatzkin, E. Lanza, D. Corle, et al., "Lack of effect of a low-fat, high-fiber diet on the recurrence of colorectal adenomas," *N Engl J Med* (2000) 342:1149–55.

161. IOM, *Dietary Reference Intakes (Macronutrients)*, pp. 377–80; and M.D. Holmes, S. Liu, S.E. Hankinson, et al., "Dietary carbohydrates, fiber, and breast cancer risk," *Am J Epidemiol* (2004) 159:732–39.

162. IOM, *Dietary Reference Intakes (Macronutrients)*, pp. 362–69.

163. E.B. Rimm, A. Ascherio, E. Giovannucci, et al., "Vegetable, fruit, and cereal fiber intake and risk of coronary heart disease among men," *JAMA* (1996) 275:447–51.

164. L.A. Bazzano, J. He, L.G. Ogden, et al., "Dietary fiber intake and reduced risk of coronary heart disease in US men and women," *Arch Intern Med* (2003) 163:1897–904.

165. A. Wolk, J.E. Manson, M.J. Stampfer, et al., "Long-term intake of dietary fiber and decreased risk of coronary heart disease among women," *JAMA* (1999) 281:1998–2004.

166. M.A. Pereira, E. O'Reily, K. Augustsson, et al., "Dietary fiber and risk of coronary heart disease: a pooled analysis of cohort studies," *Arch Intern Med* (2004) 164:370–76.

167. D.S. Ludwig, M.A. Pereira, C.H. Kroenke, et al., "Dietary fiber, weight gain, and cardiovascular disease risk factors in young adults," *JAMA* (1999) 282:1539–46.

168. P. Insel, R.E. Turner, and D. Ross, *Nutrition*, 2nd ed. (Sudbury, MA: Jones and Bartlett, 2004).

169. IOM, *Dietary Reference Intakes (Macronutrients)*, p. 348.

170. WHO/FAO, *Diet, Nutrition*; M.A. Pereira and D.S. Ludwig, "Dietary fiber and body-weight regulation: observations and mechanisms," *Pediatr Clin North Am* (2001) 48:969–80; and N.C. Howarth, E. Saltzman, and S.B. Roberts, "Dietary fiber and weight regulation," *Nutr Rev* (2001) 59:129–39.

171. J.H. Cummings, "The effect of dietary fiber on fecal weight and composition," in G.A. Spiller, ed., *CRC Handbook of Dietary Fiber in Human Nutrition* (Boca Raton: CRC Press, 1993), pp. 263–349; cited in IOM, *Dietary Reference Intakes (Macronutrients)*, p. 371.

172. IOM, *Dietary Reference Intakes (Macronutrients)*, pp. 389, 1036.

173. E.H. Haddad, L.S. Berk, J.D. Kettering, et al., "Dietary intake and biochemical, hematologic, and immune status of vegans compared with non-vegetarians," *Am J Clin Nutr* (1999) 70(suppl):586S–93S; M.S. Donaldson, "Food and nutrient intakes of Hallelujah vegetarians," *Nutr Food Sci* (2001) 31:293–303; N.E. Allen, P.N. Appleby, G.K. Davey, et al., "The associations of diet with serum insulin-like growth factor I and its main binding proteins in 292 women meat-eaters, vegetarians, and vegans," *Cancer Epidemiol Biomarkers Prev* (2002) 11:1441–48; Appleby, Davey, and Key, "Hypertension"; Davey et al., "EPIC-Oxford"; and E.A. Spencer, P.N. Appleby, G.K. Davey, et al., "Diet and body mass index in 38,000 EPIC-Oxford meat-eaters, fish-eaters, vegetarians and vegans," *Int J Obes Relat Metab Disord* (2003) 27:728–34.

174. IOM, *Dietary Reference Intakes for Thiamin, Riboflavin, Niacin, Vitamin B6, Folate, Vitamin B12, Pantothenic Acid, Biotin, and Choline* (Washington, DC: National Academies Press, 1998), pp. 260–66.

175. IOM, *Dietary Reference Intakes for Thiamin*, pp. 265, 269; and Cotton et al., "Dietary sources of nutrients."

176. E.P. Quinlivan and J.F. Gregory III, "Effect of food fortification on folic acid intake in the United States," *Am J Clin Nutr* (2003) 77:221–25.

177. N.S. Green, "Folic acid supplementation and prevention of birth defects," *J Nutr* (2002) 132(suppl):2356S–60S.

178. Haddad et al., "Dietary intake."

179. Cotton et al., "Dietary sources of nutrients."

180. Davey et al., "EPIC-Oxford"; and Toohey et al., "Cardiovascular disease risk factors."

181. IOM, *Dietary Reference Intakes for Water, Potassium, Sodium, Chloride, and Sulfate* (Washington, DC: National Academies Press, 2004), p. 244.

182. Cotton et al., "Dietary sources of nutrients."

183. IOM, *Dietary Reference Intakes for Water*, pp. 200–19; and P.K. Whelton, J. He, J.A. Cutler, et al., "Effects of oral potassium on blood pressure," *JAMA* (1997) 277:1624–32.

184. K.L. Tucker, M.T. Hannan, H. Chen, et al., "Potassium and fruit and vegetables are associated with greater bone mineral density in elderly men and women," *Am J Clin Nutr* (1999) 69:727–36; H.M. MacDonald, S.A. New, M.H. Golden, et al., "Nutritional associations with bone loss during the menopausal transition: evidence of a beneficial effect of calcium, alcohol, and fruit and vegetable nutrients and of a detrimental effect of fatty acids," *Am J Clin Nutr* (2004) 79:155–65; S.A. New, H.M. MacDonald, M.K. Campbell, et al., "Lower estimates of net endogenous non-carbonic acid production are positively associated with indexes of bone health in premenopausal and perimenopausal women," *Am J Clin Nutr* (2004) 79:131–38; and IOM, *Dietary Reference Intakes for Water*, pp. 219–22.

185. G.C. Curhan, W.C. Willett, E.R. Rimm, et al., "A prospective study of dietary calcium and other nutrients and the risk of symptomatic kidney stones," *N Engl J Med* (1993) 328:833–38; G.C. Curhan, W.C. Willett, F.E. Speizer, et al., "Comparison of dietary calcium with supplemental calcium and other nutrients as factors affecting the risk of kidney stones in women," *Ann Intern Med* (1997) 126:497–504; T. Hirvonen, P. Pietinen, M. Virtanen, et al., "Nutrient intake and use of beverages and risk of kidney stones among male smokers," *Am J Epidemiol* (1999) 150:187–94; and P. Barcelo, O. Wuhl, E. Servitge, et al., "Randomized double-blind study of potassium citrate in idiopathic hypocitraturic calcium nephrolithiasis," *J Urol* (1993) 150:1761–64.

186. P.M. Kris-Etherton, "AHA Scientific Advisory: monounsaturated fatty acids and risk of cardiovascular disease," *Circulation* (1990) 100:1253–58; and F.B. Hu, M.J. Stampfer, J.E. Manson, et al., "Dietary fat intake and the risk of coronary heart disease in women," *N Engl J Med* (1997) 337:1491–99.

187. Hu et al., "Dietary fat intake."

188. USDA ERS, Food availability database.

189. B.C. Davis and P.M. Kris-Etherton, "Achieving optimal essential fatty acid status in vegetarians: current knowledge and practice implications," *Am J Clin Nutr* (2003) 78(suppl):640S–46S; and WHO/FAO, *Diet, Nutrition*.

190. IOM, *Dietary Reference Intakes (Macronutrients)*, p. 1324; and American Dietetic Association and Dietitians of Canada. Position of the American Dietetic Association and Dietitians of Canada: vegetarian diets. *J Am Diet Assoc* (2003) 103:748–65.

191. IOM, *Dietary Reference Intakes for Vitamin C, Vitamin E, Selenium, and Carotenoids* (Washington, DC: National Academies Press, 2000), p. 52.

192. Alpha-Tocopherol, Beta Carotene Cancer Prevention Study Group, "The effect of vitamin E and beta carotene on the incidence of lung cancer and other cancers in male smokers," *N Engl J Med* (1994) 330:1029–35.

193. A.J. Duffield-Lillico, B.L. Dalkin, M.E. Reid, et al., "Selenium supplementation, baseline plasma selenium status and incidence of prostate cancer: an analysis of the complete treatment period of the Nutritional Prevention of Cancer Trial," *BJU Int* (2003) 91:608–12; L.C. Clark, B. Dalkin, A. Krongrad, et al., "Decreased incidence of prostate cancer with

selenium supplementation: results of a double-blind cancer prevention trial," *Br J Urol* (1998) 81:730–34; and G.F. Combs Jr., "Status of selenium in prostate cancer prevention," *Br J Cancer* (2004) 91:195–99.

194. R.H. Liu, "Health benefits of fruit and vegetables are from additive and synergistic combinations of phytochemicals," *Am J Clin Nutr* (2003) 78(suppl):517S–20S.

195. W.J. Craig, "Health-promoting phytochemicals: beyond the traditional nutrients," in J. Sabate, ed., *Vegetarian Nutrition* (Boca Raton: CRC Press, 2001).

196. K. Wakabayashi, M. Nagao, H. Esumi, et al., "Food-derived mutagens and carcinogens," *Cancer Res* (1992) 52:2092S–98S; and E.G. Snyderwine, "Some perspective on the nutritional aspects of breast cancer research: food derived heterocyclic amines as etiologic agents in human mammary cancer," *Cancer* (1994) 74:1070–77.

197. N. Kazerouni, R. Sinha, C.-H. Hsu, et al., "Analysis of 200 food items for benzo[a]pyrene and estimation of its intake in an epidemiologic study," *Food Chem Toxicol* (2001) 39:423–36.

198. World Cancer Research Fund/American Institute for Cancer Research, *Food, Nutrition, and the Prevention of Cancer: A Global Perspective* (Washington, DC: American Institute for Cancer Research, 1997).

199. Public Health Service, National Toxicology Program, *Report on Carcinogens*, 11th ed. (Washington, DC: DHHS, 2004).

200. USDA NASS, *Agricultural Chemical Usage: 2001 Field Crops Summary* (2002), http://usda. mannlib.cornell.edu/reports/nassr/other/pcu-bb/agcs0502.pdf, pp. 6, 68; and USDA NASS, *Agricultural Statistics 2003* (2003), www.usda.gov/nass/pubs/agr03/03_ch1.pdf, pp. 1–47.

201. For example, the International Agency for Research on Cancer publishes monographs that include reviews of several pesticides that have been linked to carcinogenicity; see http://monographs.iarc.fr/ENG/Monographs/allmonos90.php.

202. Extension Toxicology Network, "Pesticide information profiles," http://extoxnet.orst. edu/pips/ghindex.html, accessed Mar. 26, 2004; and A. Blair, "An overview of potential health hazards among farmers from use of pesticides," paper presented at Surgeon General's Conference on Agricultural Safety and Health, Des Moines, Apr. 30–May 3, 1991 (Cincinnati: National Institute of Occupational Safety and Health, 1992), p. 237.

203. Extension Toxicology Network, "Pesticide information profiles."

204. Blair, "Overview."

205. A. Blair, "Clues to cancer etiology from studies of farmers," *Scand J Environ Health* (1992) 18:209–15.

206. P.K. Mills and R. Yang, "Prostate cancer risk in California farm workers," *J Occup Environ Med* (2003) 45:249–58.

207. M.C. Alavanja, C. Samanic, M. Dosemeci, et al., "Use of agricultural pesticides and prostate cancer risk in the Agricultural Health Study cohort," *Am J Epidemiol* (2003) 157:800–14.

208. S.H. Zahm, D.D. Weisenburger, R.C. Saal, et al., "A case-control study of non-Hodgkin's lymphoma and the herbicide 2,4-dichlorophenoxyacetic acid (2,4-D) in eastern Nebraska," *Epidemiol* (1990) 1:349–56.

209. From USDA NASS, Agricultural chemical use database (Arlington, VA: National Science Foundation Center for Integrated Pest Management, 2004), www.pestmanagement.info/ nass/app_usageExcel.cfm.

210. J.A. Rusiecki, A. De Roos, W.J. Lee, et al., "Cancer incidence among pesticide applicators exposed to atrazine in the Agricultural Health Study," *J Natl Cancer Inst* (2004) 96:1375–82; and A.J. De Roos, A. Blair, J.A. Rusiecki, et al., "Cancer incidence among glyphosate-

exposed pesticide applicators in the Agricultural Health Study," *Environ Health Perspect* (2005) 113:49–54.

211. Centers for Disease Control and Prevention (CDC), *National Report on Human Exposure to Environmental Chemicals* (Atlanta, 2001), p. 35; and EPA, "Organophosphate pesticides in food—a primer on reassessment of residue limits" (1999), www.epa.gov/pesticides/op/primer.htm.

212. USDA NASS, *2001 Field Crops Summary*; USDA NASS, *Agricultural Statistics 2003*; and National Center for Food and Agricultural Policy, National pesticide use database, www.ncfap.org/database/download/database.xls, accessed Jan. 30, 2006.

213. J.E. Davies, "Neurotoxic concerns of human pesticide exposures," *Am J Ind Med* (1990) 18:327–31; and L. Rosenstock, W. Daniell, S. Barnhart, et al., "Chronic neuropsychological sequelae of occupational exposure to organophosphate insecticides," *Am J Ind Med* (1990) 18:321–25.

214. J.A. Thomas, "Toxic responses of the reproductive system," in C.D. Klassen, ed., *Casarett & Doull's Toxicology*, 5th ed. (New York: McGraw-Hill, 1996), pp. 547–82.

215. National Research Council, *Pesticides in the Diets of Infants and Children* (Washington, DC: National Academies Press, 1993); and D. Pimentel and A. Greiner, "Environmental and socioeconomic costs of pesticide use," in D. Pimentel, ed., *Techniques for Reducing Pesticide Use* (New York: John Wiley & Sons, 1997), p. 54.

216. R. Repetto and S.S. Baliga, *Pesticides and the Immune System: The Public Health Risks* (Washington, DC: World Resources Institute, 1996).

217. Public Health Service, *Carcinogens*.

218. Environmental Working Group, *PCBs in Farmed Salmon: Factory Methods, Unusual Results* (2003), www.ewg.org/reports/farmedPCBs/es.php.

219. D. Schardt, "Farmed salmon under fire," *Nutrition Action Healthletter* (2004) 31(5):9–11.

220. A. Schecter, O. Papke, K. Tung, et al., "Polybrominated diphenyl ethers contamination of United States food," *Environ Sci Technol* (2005) 38:5306–11.

221. EPA, Office of Pollution Prevention and Toxics, "Polybrominated diphenylethers (PBDEs) significant new use rule (SNUR) questions and answers," www.epa.gov/oppt/pbde/pubs/qanda.htm, accessed Aug. 15, 2005.

222. EPA, "What you need to know about mercury in fish and shellfish," www.epa.gov/waterscience/fishadvice/advice.html, accessed Mar. 30, 2005; and "Mercury in tuna: new safety concerns," *Consumer Reports* July 2006.

223. R.J. Deckelbaum, E.A. Fisher, M. Winston, et al., "Summary of a scientific conference on preventive nutrition: pediatrics to geriatrics," *Circulation* (1999) 100:450–56; and ACS, "Unified dietary guidelines" (1999), www.cancer.org/docroot/NWS/content/NWS_1_1x_Unified_Dietary_Guidelines.asp.

Argument #2. Less Foodborne Illness (pp. 59-72)

1. P.S. Mead, L. Slutsker, V. Dietz, et al., "Food-related illness and death in the United States," *Emerging Infectious Diseases* (1999) 5:607–25; and U.S. Department of Agriculture, Economic Research Service (USDA ERS), *Economics of Food-borne Disease* (Washington, DC: Government Printing Office, 2003).

2. Mead et al., "Food-related illness."

3. Center for Science in the Public Interest (CSPI), *Outbreak Alert!* (2005), www.cspinet.org/foodsafety/outbreak_report.html.

4. C. Sugarman, "Rising fears over food safety: battling the hidden hazard of bacterial contamination," *Washington Post* July 23, 1986:E1; and J. Ackerman, "Food: how safe? How altered?," *Nat Geog* May 2002:2–50.

5. CSPI, *Outbreak Alert!*

6. "How safe is that burger?," *Consumer Reports* Nov. 2002:29.

7. "Of birds and bacteria," *Consumer Reports* Jan. 2003, www.consumerreports.org/cro/food/chicken-safety-103/overview.htm.

8. A. Hingley, *Campylobacter: Low Profile Bug Is Food Poisoning Leader* (U.S. Food and Drug Administration, 1999), www.fda.gov/fdac/features/1999/599_bug.html.

9. M.F. Jacobson and C. Smith Dewaal, "Egg safety: are there cracks in the federal food safety system?," testimony before the Senate Committee on Government Affairs, July 1, 1999, www.cspinet.org/foodsafety/egg_safety.html.

10. M. Helms, P. Vastrup, P. Gerner-Smidt, et al., "Short and long term mortality associated with foodborne bacterial gastrointestinal infections: registry-based study," *BMJ* (2003) 326:357.

11. CSPI, "Kevin's law," brochure (2001), www.cspinet.org/foodsafety/kevinslawbrochure.pdf.

12. G. Manning, "Going whole hog for farm security," *USA Today* Apr. 3, 2003:9D.

13. R. Tauxe, Centers for Disease Control and Prevention, Foodborne and Diarrheal Diseases Branch, appearing on "Modern Meat," *Frontline*, Apr. 18, 2002.

14. CSPI, *Outbreak Alert!*

15. Ackerman, "Food."

16. E.B. Solomon, S. Yaron, K.R. Matthews, et al., "Transmission of *Escherichia coli* O157: H7 from contaminated manure and irrigation water to lettuce plant tissue and its subsequent internalization," *Appl Environ Microbiol* (2002) 68:397–400; and P. Belluck and C. Drew, "Tracing bout of illness to small lettuce farm," *New York Times* Jan. 5, 1998:A1.

17. P. Brasher, "Record recalls hit meat industry," *Des Moines Register* Dec. 8, 2002:1A; T. Breuer, D.H. Benkel, R.L. Shapiro, et al., "A multistate outbreak of *Escherichia coli* O157: H7 infections linked to alfalfa sprouts grown from contaminated seeds," *Emerg Infect Dis* (2001) 7:977–82; and B. Allen, "From the editor," *Nat Geog* May 2002:1.

18. "Epidemiologic notes and reports multistate outbreak of *Salmonella poona* infections: United States and Canada, 1991," *MMWR* (1991) 40(32):549–52; and Associated Press, "Kale, turnip greens recalled," Dec. 26, 2002.

19. A. Nunez-Delgado, E. Lopez-Periago, F. Diaz-Fierros Vigueira, et al., "Chloride, sodium, potassium and faecal bacteria levels in surface runoff and subsurface percolates from grassland plots amended with cattle slurry," *Bioresour Technol* (2002) 82:261–71; U.S. Dept. of Agriculture Animal Plant Health Inspection Service (USDA APHIS), *Cryptosporidium and Giardia in Beef Calves* (Fort Collins, CO, 2001), http://nahms.aphis.usda.gov/beefcowcalf/chapa/ChapaCrypto.pdf; and I.V. Wesley, S.J. Wells, K.M. Harmon, et al., "Fecal shedding of *Campylobacter* and *Arcobacter* spp. in dairy cattle," *Appl Environ Microbiol* (2000) 66:1994–2000.

20. USDA APHIS, *Info Sheet: Salmonella in United States Feedlots* (2001), www.aphis.usda.gov/vs/ceah/ncahs/nahms/feedlot/feedlot99/FD99salmonella.pdf; and USDA APHIS, *Info Sheet: Escherichia coli O157 in United States Feedlots* (2001), www.aphis.usda.gov/vs/ceah/ncahs/nahms/feedlot/feedlot99/FD99ecoli.pdf.

21. S.H. Lee, D.A. Levy, G.F. Craun, et al., "Surveillance for waterborne-disease outbreaks: United States, 1999–2000," *MMWR* (2002) 51:1–47.

22. I.D. Ogden, D.R. Fenlon, A.J. Vinten, et al., "The fate of *Escherichia coli* O157 in soil and its potential to contaminate drinking water," *Int J Food Microbiol* (2001) 66:111–7; and

S.G. Jackson, R.B. Goodbrand, R.P. Johnson, et al., "*Escherichia coli* O157:H7 diarrhoea associated with well water and infected cattle on an Ontario farm," *Epidemiol Infect* (1998) 120:17–20.

23. Lee et al., "Surveillance."

24. A.J. Lung, C.M. Lin, J.M. Kim, et al., "Destruction of *Escherichia coli* O157:H7 and *Salmonella enteritidis* in cow manure composting," *J Food Prot* (2001) 64:1309–14.

25. I.T. Kudva, K. Blanch, and C.J. Hovde, "Analysis of *Escherichia coli* O157:H7 survival in ovine or bovine manure and manure slurry," *Appl Environ Microbiol* (1998) 64:3166–74.

26. E.E. Natvig, S.C. Ingham, B.H. Ingham, et al., "*Salmonella enterica* serovar typhimurium and *Escherichia coli* contamination of root and leaf vegetables grown in soils with incorporated bovine manure," *Appl Environ Microbiol* (2002) 68:2737–44.

27. Kudva, Blanch, and Hovde, "*Escherichia coli* O157:H7 survival."

28. M.E. Ensminger, *Animal Science*, 9th ed. (Danville, IL: Interstate Publishing, 1991), pp. 31–32.

29. L. Saif, "Panel dialogue: challenges faced and met in research on food health," National Academies Workshop, Exploring a Vision: Integrating Knowledge for Food and Health, June 9, 2003, Washington, DC.

30. J.A. Zahn, "Evidence for transfer of tylosin and tylosin-resistant bacteria in air from swine production facilities using sub-therapeutic concentrations of tylan in feed," presentation at International Animal Agriculture and Food Science Conference, July 24–28, 2001, Indianapolis.

31. B.Z. Predicala, J.E. Urban, R.G. Maghirang, et al., "Assessment of bioaerosols in swine barns by filtration and impaction," *Curr Microbiol* (2002) 44:136–40.

32. Appearing on *Morning Edition*, Oct. 10, 2005, National Public Radio, www.npr.org.

33. D.J. Alexander and I.H. Brown, "Recent zoonoses caused by influenza A viruses," *Rev Sci Tech* (2000) 19:197–225; Centers for Disease Control and Prevention, National Center for Infectious Diseases (CDC NCID), *The Influenza (Flu) Viruses* (2003), www.cdc.gov/ncidod/diseases/flu/viruses.htm; and World Health Organization (WHO), *Avian Influenza: Assessing the Pandemic Threat* (2005), www.who.int/csr/disease/influenza/H5N1-9reduit.pdf.

34. CDC NCID, *Influenza (Flu) Viruses.*

35. G. Kolata, *Flu: The Story of the Great Influenza Pandemic* (Darby, PA: Diane Publishing Co, 2001).

36. U.S. Department of Health and Human Services (DHHS), "What is an influenza pandemic?," www.pandemicflu.gov/general/whatis.html, accessed June 4, 2006; Alexander and Brown, "Recent zoonoses"; CDC, "Information about influenza epidemics," www.cdc.gov/flu/avian/gen-info/pandemics.htm, accessed Nov. 22, 2005; and R. Stein, "Infections now more widespread," *Washington Post* June 15, 2003:A1.

37. T.K. Taubenberger, A.H. Reid, R.M. Lourens, et al., "Characterization of the 1918 influenza virus polymerase genes," *Nature* (2005) 437:889–93; and L.K. Altman, "New microbes could become new norm," *New York Times* Mar. 9, 2004:D6.

38. B.W.J. Mahy and C.C. Brown, "Emerging zoonoses: crossing the species barrier," *Rev Sci Tech* (2000) 19:33–40; J. Taylor, "Hong Kong watching for bird flu," *Australian Broadcasting News* Feb. 2, 2004, www.abc.net.au/pm/content/2004/s1036587.htm; B. Wuethrich, "Chasing the fickle swine flu," *Science* (2003) 299:1502–05; Health, Welfare, and Food Bureau of Hong Kong, "Preventive and contingency measures to combat avian influenza in Hong Kong" (2004), www.info.gov.hk/info/flu/eng/files/legco-e.pdf.

39. M. Mellon, C. Benbrook, and K. Benbrook, *Hogging It* (Cambridge, MA: UCS Publications, 2001).

40. M. Swartz, "Human diseases caused by foodborne pathogens of animal origin," *Clin Infect Dis* (2002) 34(3):S111–22; and S.B. Levy, G.B. FitzGerald, and A.B. Macone, "Changes in intestinal flora of farm personnel after introduction of a tetracycline-supplemented feed on a farm," *New Engl J Med* (1976) 295:583–88.

41. U.S. Government Accountability Office, *Antibiotic Resistance: Federal Agencies Need to Better Focus Efforts to Address Risk to Humans from Antibiotic Use in Animals*, Report No. GAO-04-490 (2004), www.gao.gov/new.items/d04490.pdf, appendix VII: Comments from the Department of Health and Human Services.

42. M. Barza and K. Travers, "Excess infections due to antimicrobial resistance: the attributable fraction," *Clin Infect Dis* (2002) 34(3):S126–30.

43. D.G. White, S. Zhao, R. Sudler, et al., "The isolation of antibiotic-resistant *Salmonella* from retail ground meats," *New Engl J Med* (2001) 345:1147–53.

44. S.D. Holmberg, M.T. Osterholm, K.A. Senger, et al., "Drug-resistant *Salmonella* from animals fed antimicrobials," *New Engl J Med* (1984) 311(10):617–22; also see T.F. O'Brien, J.D. Hopkins, E.S. Gilleece, et al., "Molecular epidemiology of antibiotic resistance in *Salmonella* from animals and human beings in the United States," *New Engl J Med* (1982) 307(1):1–6.

45. CDC, *Human Isolates Final Report, 2002: National Antimicrobial Resistance Monitoring System for Enteric Bacteria (NARMS)* (2004), www.cdc.gov/narms/annual/2002/ 2002ANNUALREPORTFINAL.pdf.

46. U.S. Food and Drug Administration, "Enroflaxin for poultry; opportunity for hearing," Docket No. 00N-1571, *Fed Reg* (2000) 65(211):64954–65, www.fda.gov/OHRMS/DOCKETS/ 98fr/103100a.htm.

47. FAAIR Scientific Advisory Panel, "Select findings and conclusions," *Clin Infect Dis* (2002) 34(Suppl 3):S73–75.

48. A.W. Mathews and Z. Goldfarb, "FDA bans use of antibiotic in poultry," *Wall Street Journal* July 29, 2005:B1.

49. Animal Health Institute, "The antibiotics debate" (2004), www.ahi.org/antibioticsDebate/ index.asp, accessed May 2, 2005.

50. WHO, *Global Principles for the Containment of Antimicrobial Resistance in Animals Intended for Food* (Geneva, 2001), http://whqlibdoc.who.int/hq/2000/WHO_CDS_ CSR_APH_2000.4.pdf; K.M. Shea and the Committee on Environmental Health and the Committee on Infectious Diseases, "Nontherapeutic uses of antimicrobial agents in animal agriculture: implications for pediatrics," *Pediatrics* (2004) 114(3):862–68; Institute of Medicine, *Microbial Threats to Human Health: Emergence, Detection, and Response* (Washington, DC: National Academies Press, 2003), p. 208; see www. keepantibioticsworking.org for in-depth information about antibiotic resistance. The Preservation of Antibiotics for Medical Treatment Act of 2005 is S.742 and H.R. 2562.

51. National Research Council, *The Use of Drugs in Food Animals, Benefits and Risks* (Washington, DC: National Academies Press, 2002), p. 157.

52. European Union, "Ban on antibiotics as growth promoters in animal feed enters into effect," press release, Dec. 22, 2005.

53. J. Callesen, "Effects of termination of AGP use on pig welfare and productivity," in WHO, *Working Papers from the International Invitational Symposium: Beyond Antimicrobial Growth Promoters in Food Animal Production* (Nov. 6–9, 2002, Foulum, Denmark), www. agrsci.dk/djfpublikation/djfpdf/djfhu57.pdf.

54. WHO, *Working Papers*, www.agrsci.dk/djfpublikation/djfpdf/djfhu57.pdf, p. 18; F.M. Aarestrup, "Effect of abolishment of the use of antimicrobial agents for growth promotion on occurrence of antimicrobial resistance in fecal enterococci from food animals in Denmark," *Antimicrob Agents Chemother* (2001) 45:2056–59; and M.C. Evans

and H.C. Wegener, "Antimicrobial growth promoters and *Salmonella* spp., *Campylobacter* spp. in poultry and swine, Denmark," *Emerg Infect Dis* (2003) 9(4):489–92.

55. Danish Institute for Food and Veterinary Research, *DANMAP 2004: Use of Antimicrobial Agents and Occurrence of Antimicrobial Resistance in Bacteria from Food Animals, Foods and Humans in Denmark* (2005), www.dfvf.dk/Files/Filer/Zoonosecentret/Publikationer/Danmap/Danmap_2004.pdf, figure 2; and WHO, *Impacts of Antimicrobial Growth Promoter Termination in Denmark: The WHO International Review Panel's Evaluation of the Termination of the Use of Antimicrobial Growth Promoters in Denmark* (2003), www.who.int/salmsurv/en/Expertsreportgrowthpromoterdenmark.pdf, pp. 41–44.

56. M. Burros, "Poultry industry quietly cuts back on antibiotic use," *New York Times* Feb. 10, 2002:A1; E. Weise, "'Natural' chickens take flight: four top producers end use of antibiotics," *USA Today* Jan. 24, 2006:5D; and Iowa Pork Producers Association, "Proposed resolution number 2004–5: feeding of growth promotant antibiotics" (2004), www.iowapork.org/download/2004_resolutions.pdf.

57. Animal Health Institute, "Antibiotic use in animals rises in 2004," news release, June 27, 2005, www.ahi.org/mediaCenter/documents/Antibioticuse2004.pdf.

58. D. Schuettler, "Scientists fear bird flu could trigger pandemic: global action must be taken immediately, conference told," *National Post* (Reuters) Feb 24, 2005:A16.

59. CDC, "Recent avian influenza outbreaks in Asia," www.cdc.gov/flu/avian/outbreaks/asia.htm, accessed Mar. 3, 2005.

60. S. Leahy, "Bird flu defeated—at high cost," Inter Press News Service Agency, Aug. 27, 2004, www.ipsnews.net/interna.asp?idnews=25254.

61. D. Milbank, "Capitol Hill flu briefing was no trick, and no treat," *Washington Post* Oct. 13, 2005:A2.

62. CSPI, *Outbreak Alert!*

Argument #3. Better Soil (pp. 73-85)

1. C. Niskanen, "Trout in troubled waters: shifts in land use in southeast Minnesota are causing sediment damage to streams," *St. Paul Pioneer Press* Apr. 17, 2005:7G.

2. U.S. Department of Agriculture, Economic Research Service (USDA ERS), "Briefing room: land use, value, and management: major uses of land" (2002), www.ers.usda.gov/Briefing/LandUse/majorlandusechapter.htm, accessed May 2, 2003.

3. G. Wuerthner, freelance biologist and former employee of U.S. Bureau of Land Management, email to Center for Science in the Public Interest (CSPI), Sept. 16, 2004.

4. U.S. Department of Agriculture, Natural Resources Conservation Service (USDA NRCS), *National Resources Inventory 2001 NRI: Soil Erosion* (2003), www.nrcs.usda.gov/technical/land/nri01/erosion.pdf.

5. USDA ERS, *Agricultural Resources and Environmental Indicators* (Washington, DC, 2003), p. 4.2-15.

6. M. Al-Kaisi, "Soil erosion and crop productivity: topsoil thickness" (Ames, IA: Iowa State University, 2001), www.ipm.iastate.edu/ipm/icm/2001/1-29-2001/topsoilerosion.html.

7. The 37 percent figure is from United Nations Development Programme, United Nations Environment Programme, World Bank, and World Resources Institute, *World Resources 2000–2001: People and Ecosystems—The Fraying Web of Life* (Washington, DC, 2001), pp. 258–59.

8. USDA ERS, *Agricultural Resources*, pp. 4.2-14, 15.

9. USDA ERS, *Agricultural Resources*, pp. 4.2-14, 15.

10. USDA ERS, *Soil, Nutrient and Water Management Systems Used in U.S. Corn Production* (Washington, DC, 2002), p. 9.

11. USDA ERS, *Agricultural Resources*, pp. 4.2-14, 15.

12. USDA ERS, *Summary Report 1997 National Resources Inventory* (Washington, DC, 2000), pp. 51, 57. Much of the data on soil erosion in this chapter are adapted from that report. Although the 2002 *Inventory* has been published, it is not as exhaustive as the 1997 report, and USDA maintains that data from the 1997 report are more reliable and consistent. For further explanation, see www.nrcs.usda.gov/technical/NRI/.

13. USDA ERS, *Summary Report*, pp. 58–95. USDA NRCS estimates that water erosion impairs crop productivity on about 65 million acres, and wind erosion impairs productivity on 48 million acres. Some of that land experiences both types of erosion. Current national data do not allow distinguishing the extent of erosion related to different crops. If those data were available, one could estimate the erosion resulting from animal agriculture.

14. USDA NRCS, *Managing Soil Organic Matter: The Key to Air and Water Quality* (2003), www.nm.nrcs.usda.gov/technical/tech-notes/soils/soil2.pdf.

15. P. Sullivan, *Overview of Cover Crops and Green Manures: Fundamentals of Sustainable Agriculture* (National Sustainable Agriculture Information Service, 2003), http://attra.ncat.org/attra-pub/PDF/covercrop.pdf.

16. USDA ERS, *Summary Report*, pp. 58–59.

17. USDA NRCS, "What is topsoil worth?," http://soils.usda.gov/sqi/concepts/soil_organic_matter/som_d.html, accessed Dec. 26, 2005.

18. W.R. Osterkamp, hydrologist, U.S. Geological Survey, email to CSPI, Apr. 25, 2003.

19. USDA NRCS, *Managing Soil Organic Matter*.

20. A. Fletcher, "Soil erosion could devastate food sector" (2006), www.foodnavigator.com/news/ng.asp?n=66605-soil-nutrients-crops.

21. USDA Agricultural Research Service (USDA ARS), "Technologies for management of arid rangelands," research project description (2001), www.ars.usda.gov/research/publications/Publications.htm?seq_no_115=142788; J. Daniel, Grazinglands Research Laboratory, USDA ARS, email to CSPI, Apr. 28, 2003; and J.A. Daniel and W.A. Phillips, "Impacts of grazing strategies on soil compaction," paper presented at American Society of Agricultural Engineers 2000 Summer Meeting, Milwaukee, July 9–12, 2000.

22. A.J. Jones, R.D. Grisso, and C.A. Shapiro, "Soil compaction … fact and fiction: common questions and their answers" (Lincoln: University of Nebraska Cooperative Extension Service, 1988), http://ianrpubs.unl.edu/soil/cc342.htm.

23. J.A. Daniel, P. Kenneth, W. Altom, et al., "Long-term grazing density impacts on soil compaction," *Trans ASAE* (2002) 45:1911–15.

24. U.S. Geological Survey, "An introduction to biological soil crusts," www.soilcrust.org/crust101.htm, accessed June 17, 2004.

25. J. Belsky and J.L. Gelbard, "Comrades in harm: livestock and weeds in the intermountain west," in G. Wuerthner and M Matteson, eds., *Welfare Ranching: The Subsidized Destruction of the American West* (Washington, DC: Island Press, 2002), pp. 203–06.

26. USDA NRCS, *Summary Report 1997 National Resources Inventory* (Washington, DC, 2000), p. 9. For similar 2003 data, see USDA, "Johanns announces 43 percent decline in total cropland erosion," press release, May 22, 2006, www.usda.gov/2006/05/0170.xml.

27. USDA Farm Service Agency, Conservation Reserve Program monthly contract report, www.fsa.usda.gov/crpstorpt/06Approved/r1sumyr/us.htm, accessed Aug. 3, 2005.

28. USDA NRCS, *National Resources Inventory: 2002* (Washington, DC, 2004), p. 1.

29. USDA, *Agricultural Resources and Environmental Indicators* (Washington, DC, 2003), ch. 4.2, pp. 22, 41.

30. Purdue University, "Tillage type definitions" (2002), www.ctic.purdue.edu/Core4/CT/Definitions.html.

31. Calculations based on acreages in USDA, National Agricultural Statistics Service (USDA NASS), *Agricultural Chemical Usage: Field Crops Summary* for 1998, 2000, 2001; and grain used for feed from *Agricultural Outlook* Sept. 2002, www.ers.usda.gov/publications/agoutlook/sep2002/ao294.pdf, p. 44, table 17.

32. Calculations based on acreages in USDA NASS, *Field Crops Summary* for 1998, 2000, 2001. Total U.S. fertilizer use in 2001 was 20.6 million tons according to USDA ERS, "Agricultural chemicals and production technology: questions and answers, 2002," ERS Online Briefing Room, www.ers.usda.gov/Briefing/AgChemicals/Questions/nmqa2.htm, accessed Mar. 23, 2004.

33. C.E. Pitcairn, U.M. Skiba, M.A. Sutton, et al., "Defining the spatial impacts of poultry farm ammonia emissions on species composition of adjacent woodland groundflora using Ellenberg Nitrogen Index, nitrous oxide and nitric oxide emissions and foliar nitrogen as marker variables," *Environ Pollut* (2002) 119:9–21.

34. Adapted from USDA NASS, *Milk Production, Disposition, and Income 2002 Summary* (Washington, DC, 2003), p. 2; USDA NASS, *Poultry Slaughter 2002 Summary* (Washington, DC, 2003), p. 2; USDA NASS, *Livestock Slaughter 2002 Summary* (Washington, DC, 2003), pp. 35, 41, 49; and USDA NASS, *Chickens and Eggs 2003 Summary* (Washington, DC, 2004), p. 2.

35. National Research Council (NRC), *Air Emissions from Animal Feeding Operations: Current Knowledge, Future Needs* (Washington, DC: National Academies Press, 2003), ch. 3.

36. United Nations Industrial Development Organization, *Technical Report No. 26 Part 1: Mineral Fertilizer Production and the Environment* (Geneva, 1998), p. 49; and NRC, *Air Emissions*, p. 75.

37. Potash and Phosphate Institute and Potash and Phosphate Institute of Canada (PPI-PPIC), *Technical Bulletin 2002–1: Plant Nutrient Use in North American Agriculture— Producing Food and Fiber, Preserving the Environment, Integrating Organic and Inorganic Sources* (Norcross, GA, 2002), p. 60.

38. D. Eckert, "Efficient fertilizer use: fertilizer management practices" (Bannockburn, IL: IMC-Agrico), www.agcentral.com/imcdemo/05Nitrogen/05-0.htm; and A. Napgezek, "Aging soils?," *University of Wisconsin Extension NPM Field Notes* Feb./Mar. 1999.

39. H. de Zeeuw and K. Lock, "Urban and periurban agriculture, health and environment," discussion paper for Food and Agriculture Organization of the United Nations-Resource Centre for Urban Agriculture and Forestry electronic conference, Urban and Periurban Agriculture on the Policy Agenda (2000), www.fao.org/urbanag/Paper2-e.htm.

40. U.S. Environmental Protection Agency, Office of Pollution Prevention and Toxics (EPA OPPT), *Background Report on Fertilizer Use, Contaminants, and Regulations* (1999), www.epa.gov/opptintr/fertilizer.pdf, pp. ii, iv; and U. Krogmann and L.S. Boyles, *Land Application of Sewage Sludge (Biosolids), No. 5: Heavy Metals* (New Brunswick, NJ: Rutgers University Agricultural Experiment Station, 1999).

41. EPA OPPT, *Background Report*, p. 112.

42. EPA OPPT, *Background Report*, p. 110.

43. J. Kaplan, Z. Ross, and B. Walker, *As You Sow: Toxic Waste in California Home and Farm Fertilizers* (San Francisco: California Public Interest Research Group, 1999), p. 1.

44. PPI-PPIC, *Technical Bulletin*, p. 48.

45. R.L. Wershaw, J.R. Garbarino, and M.R. Burkhardt, "Roxarsone in natural water systems," in U.S. Geological Survey, *Proceedings: Effects of Confined Animal Feeding Operations (CAFOs) on Hydrologic Resources and the Environment*, meeting held in Fort Collins, CO, Aug. 30–Sept. 1, 1999, http://water.usgs.gov/owq/AFO/proceedings/afo/html/wershaw.html.

46. Based on manure data in R.L. Kellogg, C.H. Lander, D.H. Moffitt, et al., *Manure Nutrients Relative to the Capacity of Cropland and Pastureland to Assimilate Nutrients: Spatial and Temporal Trends for the United States* (Washington, DC: USDA, 2000), p. 49; and USDA NASS data on numbers of livestock, www.nass.usda.gov:8080/QuickStats/indexbysubject. jsp?Pass_group=Livestock+%26+Animals.

47. Based on a midyear population of 285,317,559 from the U.S. Census Bureau, "State population estimates: April 1, 2000 to July 1, 2002," www.census.gov/popest/archives/ 2000s/vintage_2002/ST-EST2002-01.html, accessed Jan. 13, 2003; and an average waste generation of about 0.518 tons per person per year from EPA, *National Pollutant Discharge Elimination System Permit Regulation and Effluent Limitation Guidelines and Standards for Concentrated Animal Feeding Operations (CAFOs)*, as cited in *Fed Reg* (2003) 68(29):7175–274 (complete document is at www.epa.gov/EPA-WATER/2003/February/Day-12/w3074.htm).

48. Adapted from American Society of Agricultural Engineers, *Manure Production and Characteristics* (St. Josephs, MI, 2002), pp. 687–89; and Kellogg et al., *Manure Nutrients*, p. 49.

49. Kellogg et al., *Manure Nutrients*, p. 74.

50. Council for Agricultural Science and Technology, *Storing Carbon in Agricultural Soils to Help Mitigate Global Warming*, CAST Issue Paper 14 (Washington, DC, 2000), p. 2; and Kellogg et al., *Manure Nutrients*, pp. 53, 56.

51. PPI-PPIC, *Organic or Inorganic, Which Nutrient Source Is Better for Plants?*, Enviro-briefs No. 2 (Norcross, GA, 2002).

52. University of Maryland Cooperative Extension Service, *Nutrient Manager: Making the Most of Manure* (College Park, MD, 1994).

53. K.E. Nachman, J.P. Graham, L.B. Price, and E.K. Silbergeld, "Arsenic: a roadblock to potential animal waste management solutions," *Environ Health Perspect* (2005) 113(9):1123–24.

54. J.E. Lee, "Sludge spread on fields is fodder for lawsuits," *New York Times* June 26, 2003:20.

55. EPA, Office of Enforcement and Compliance Assurance, *Land Application of Sewage Sludge: A Guide for Land Appliers on the Requirements of the Federal Standards for the Use or Disposal of Sewage Sludge, 40 CFR Part 503* (1994), www.epa.gov/owm/mtb/biosolids/sludge.pdf.

56. Lee, "Sludge."

57. EPA OPPT, *Background Report*, p. iii.

58. R. Kellogg, R. Nehring, A. Grube, et al., "Trends in the potential for environmental risk from pesticide loss from farm fields" (USDA Natural Resources Conservation Service, 1999), www.nrcs.usda.gov/technical/land/pubs/pesttrend.html.

59. Extension Toxicology Network, *Movement of Pesticides in the Environment*, Toxicology Information Brief (1993), http://extoxnet.orst.edu/tibs/movement.htm.

60. "Roundup kills frogs as well as tadpoles, Pitt biologist finds," University of Pittsburgh news release, Aug. 3, 2005, www.umc.pitt.edu:591/m/FMPro?-db=ma&-lay=a&-format=d. html&id=2115&-Find; and "Roundup highly lethal to amphibians, finds University of Pittsburgh researcher," *Medical News Today* Apr. 3, 2005, www.medicalnewstoday.com/ medicalnews.php?newsid=22159.

61. T. Hayes, K. Haston, M. Tsui, et al., "Atrazine-induced hermaphroditism at 0.1ppb in American leopard frogs (*Rana pipiens*): laboratory and field evidence," *Environ Health Perspect* (2003) 111(4):568–75; and L. Tavera-Mendoza, S. Ruby, P. Brousseau, et al., "Response of the amphibian tadpole *Xenopus laevis* to atrazine during sexual differentiation of the ovary," *Environ Toxicol Chem* (2002) 21:1264–67.

62. M. Losure, "Frog researcher invited to tell his story," Minnesota Public Radio, Oct. 26, 2004, http://news.minnesota.publicradio.org/features/2004/10/25_losurem_frogresearch/.

Argument #4. More and Cleaner Water (pp. 87-101)

1. A.J. Laukaitis, "Drought shrinking McConaughy," *Lincoln Journal Star* May 22, 2005:D1.

2. Water Education Foundation, "Colorado river project," www.water-ed.org/coloradoriver. asp.

3. Calculations based on D. Pimentel, J. Houser, E. Preiss, et al., "Water resources: agriculture, the environment, and society," *Bioscience* (1997) 47(2):97–106.

4. Calculations based on D. Pimentel, B. Berger, D. Filiberto, et al., "Water resources: agricultural and environmental issues," *Bioscience* (2004) 54:909–18.

5. U.S. Geological Survey (USGS), *Estimated Use of Water in the United States in 1995* (Washington, DC, 1998), pp. 18–19.

6. USGS, *Estimated Use of Water*, pp. 18–19; and U.S. Department of Agriculture, National Agricultural Statistics Service (USDA NASS). *2003 Farm and Ranch Irrigation Survey* (Washington, DC, 2004), pp. 69–89. Other irrigation includes vegetables and fruit orchards, irrigation of feed grains for export, other crops (e.g., rice), fish farms, parks, and public and private golf courses.

7. USDA, Economic Research Service (USDA ERS), *Agricultural Resources and Environmental Indicators* (Washington, DC, 2003) p. 2.1-1.

8. USGS, *Estimated Use of Water*, p. 19.

9. S. Postel, *Pillar of Sand* (New York: WorldWatch Institute, 1999), p. 80. The figure of 21 billion gallons per day originally was recorded by the former U.S. Water Resources Council and reported in J. Adler, "The browning of America," *Newsweek* Feb. 23, 1981:26.

10. D. Jehl, "Saving water, U.S. farmers are worried they'll parch," *New York Times* Aug. 28, 2002:A1.

11. High Plains Water Conservation District Number 1, "The Ogallala Aquifer" (Lubbock, TX), www.hpwd.com/the_ogallala.asp, accessed Mar. 29, 2004; and D. McConnell, "Groundwater: on-line resource" (University of Akron, 1998), http://lists.uakron.edu/ geology/natscigeo/Lectures/gwater/gwater.htm#ogallala, accessed Aug. 8, 2005.

12. Panhandler Plains Historical Museum, "Ogallala Aquifer" (Canyon, TX), www. panhandleplains.org/education/pop_geo_ogallala.php, accessed Mar. 11, 2005.

13. L.E. Jones, "Saltwater contamination in the Upper Floridan Aquifer at Brunswick, Georgia," in K.J. Hatcher, ed., *Proceedings of 2001 Georgia Water Resources Conference* (Athens, GA: Institute of Ecology), http://ga.water.usgs.gov/publications/gwrc2001jones. html, pp. 644–47.

14. USDA ERS, *Agricultural Resources*, p. 2.1-6.

15. National Research Council, *Mitigating Losses from Land Subsidence in the United States* (Washington, DC: National Academies Press, 1991), p. 1. The $125 million is equivalent to $180 million in 2005 dollars.

16. USDA ERS, *Agricultural Resources*, p. 2.1-2.

17. Authors' calculations based on USDA data on irrigated acreages and fractions used to feed U.S. livestock.

18. USDA NASS, *Irrigation Survey*, table 27.

19. USDA ERS, *Agricultural Resources*, p. 2.1-2.

20. USDA NASS, *Irrigation Survey*, tables 27, 28.

21. USDA NASS, *Irrigation Survey*, tables 12, 27, adjusted for fractions of irrigated crops used to feed domestic livestock.

22. USDA ERS, *Agricultural Resources*, pp. 2.2-2, 3; USDA ERS, "Briefing room: irrigation and water use" (2004), www.ers.usda.gov/Briefing/WaterUse/Questions/qa10.htm, accessed Aug. 8, 2005.

23. USDA ERS, *Agricultural Resources*, pp. 2.2-3, 7, 11.

24. S. Postel, "Growing more food with less water," *Scientific American* Feb. 2001:50.

25. USDA ERS, *Agricultural Resources*, pp. 2.2-11.

26. USDA NASS, *Irrigation Survey*, p. 2-2.11.

27. Postel, "Growing more food."

28. T.L. Anderson and P.S. Snyder, *Priming the Invisible Pump* (Bozeman, MT: Property and Environment Research Center, 1997).

29. U.S. House of Representatives, Subcommittee on Oversight and Investigations of the Committee on Insular and Interior Affairs, Committee Print, Dec. 1988; referenced in Subcommittee on Oversight and Investigations of the Committee on Natural Resources. *Taking from the Taxpayer: Public Subsidies for Natural Resource Development: An Investigative Report* (Washington, DC: U.S. Government Printing Office, 1994), pp. 41–69 (expressed in 1988 dollars).

30. U.S. House of Representatives, *Taking from the Taxpayer*, pp. 41–69; referenced in Postel, *Pillar of Sand*, p. 231.

31. U.S. House of Representatives, *Taking from the Taxpayer*, pp. 41–69; referenced in Postel, *Pillar of Sand*, p. 231.

32. Environmental Working Group, "Executive summary," in *California Water Subsidies* (2004), pp. 1–2, www.ewg.org/reports/watersubsidies/execsumm.php.

33. USDA ERS, *Agricultural Resources*, p. 2.1-2.

34. For example, 100 gallons of irrigation water increases farm income by 3.4 cents for corn and 1.6 cents for sorghum. Calculated from USDA NASS, *Irrigation Survey*, tables 27, 28; and USDA ERS, table 17, supply and utilization, *Agricultural Outlook* Jan.–Feb. 2001:37–38, www.ers.usda.gov/Publications/AgOutlook/Jan2001/ao278.pdf.

35. USDA NASS, "Quick stats," www.nass.usda.gov/Data_and_Statistics/Quick_Stats/index.asp, accessed Jan. 11, 2006; and USDA NASS, *Noncitrus Fruits and Nuts 2004 Summary* (2005), http://usda.mannlib.cornell.edu/reports/nassr/fruit/pnf-bb/ncit0705.pdf.

36. Natural Resources Defense Council, "Alfalfa: the thirstiest crop," fact sheet (2001), www.nrdc.org/water/conservation/fcawater.asp.

37. Congressional Budget Office, *Water Use Conflicts in the West: Implications for Reforming the Bureau of Reclamation's Water Supply Policies* (Washington, DC, 1997).

38. USGS, *Estimated Use of Water*, p. 37.

39. New York City Department of Environmental Protection, *New York City 2004 Drinking Water Supply and Quality Report*, www.nyc.gov/html/dep/pdf/wsstat04.pdf.

40. S.A. Ewing, D.C. Lay, and E. von Berell, *Farm Animal Well-Being: Stress Physiology, Animal Behavior, and Environmental Design* (Upper Saddle River, NJ: Prentice Hall, 1999), p. 235.

41. USGS, *Estimated Use of Water*, p. 62.

42. USDA ERS, *Confined Animal Production and Manure Nutrients* (Washington, DC, 2001), www.ers.usda.gov/publications/aib771, p. iii.

43. Yunker Plastics, Inc., "Manure lagoons," www.yunkerplastics.com/manure.htm.

44. North Carolina State University, "Frequently asked questions about livestock production," www.bae.ncsu.edu/programs/extension/manure/awm/program/barker/questions/q_doc.html, accessed Oct. 9, 2005; and P. Cantrell, "State opens gate, waterways to livestock factories," *Great Lakes Bulletin* Winter 1999:19 (Michigan Land Use Institute).

45. Associated Press, "11 million litres of liquid manure spill into upstate New York river," Aug. 13, 2005.

46. M. Cook and E. Stanley, "Reducing water pollution from animal feeding operations," testimony before the House Subcommittees on Livestock, Dairy, and Poultry and Forestry, Resource Conservation, and Research of the Committee on Agriculture, May 13, 1998, www.epa.gov/ocirpage/hearings/testimony/105_1997_1998/051398.htm.

47. D. Pimentel, C. Harvey, P. Resosudarmo, et al., "Environmental and economic costs of soil erosion and conservation benefits," *Science* (1995) 267:1117–23.

48. P.K. Koluvek, K. Tanji, and T. Trout, "Overview of soil erosion from irrigation," *J Irrig Drainage Engin* (1993) 119:929–46.

49. USDA ERS, *Agricultural Resources*, p. 2.3-5.

50. United Nations Development Programme, United Nations Environment Programme, World Bank, and World Resources Institute, *World Resources 2000–2001: People and Ecosystems—The Fraying Web of Life* (Washington, DC, 2001), p. 50.

51. Postel, *Pillar of Sand*, p. 101.

52. USDA ERS, *Agricultural Resources*, p. 2.2-1.

53. D. Neffendorf, chairman and coordinator, USDA Natural Resources Conservation Service (USDA NRCS) Grazing Land Conservation Initiative, email to Center for Science in the Public Interest (CSPI), Jan. 15, 2004; and Postel, *Pillar of Sand*, p. 93.

54. Potash and Phosphate Institute and Potash and Phosphate Institute of Canada (PPI-PPIC), *Technical Bulletin 2002 1: Plant Nutrient Use in North American Agriculture—Producing Food and Fiber, Preserving the Environment, Integrating Organic and Inorganic Sources* (Norcross, GA, 2002), p. iii.

55. Calculations based on USDA ERS, "U.S. fertilizer use and price" (1964–2003), tables 1 and 2, www.ers.usda.gov/Data/FertilizerUse/; fertilizer usage data from states. The analysis included barley, corn, oats, wheat, sorghum, soy, alfalfa, hay, and pasture, but that is not an exhaustive list, so the figure given may be an underestimate.

56. USDA ERS, "Briefing room: agricultural chemicals and production technology: nutrient management" (2005), www.ers.usda.gov/Briefing/AgChemicals/nutrientmangement. htm, accessed June 4, 2006; and USDA NASS, *Agricultural Chemical Usage 2003 Field Crops Summary* (2004), http://usda.mannlib.cornell.edu/reports/nassr/other/pcu-bb/agcs0504. pdf, p. 22.

57. PPI-PPIC, *Technical Bulletin*, p. 51.

58. N.N. Rabalais, R.E. Turner, and D. Scavia, "Beyond science into policy: Gulf of Mexico hypoxia and the Mississippi River," *BioScience* (2002) 52:129–42.

59. PPI-PPIC, *Technical Bulletin*, p. 51.

60. Rabalais, Turner, and Scavia, "Beyond science"; and National Science and Technology Council, Committee on Environment and Natural Resources (NSTC), *Integrated Assessment of Hypoxia in the Northern Gulf of Mexico* (Washington, DC, 2000), p. 3.

61. U.S. Environmental Protection Agency (EPA), "Rivers and streams," *National Water Quality Inventory 2000 Report* (2002), ch. 2, www.epa.gov/305b/2000report/chp2.pdf.

62. See, for example, G. Martin, "Phosphate risks abound," Charlotte City, FL, *Sun-Herald*, www.sun-herald.com/phosphate/part4.htm.

63. In Idaho, the sites are Eastern Michaud Flats (EPA ID IDD984666610), which was a primary processor of phosphate rock, and Kerr-McGee Chemical Corporation (EPA ID IDD041310707), which was a secondary processor of wastes from phosphate rock mining. In Florida, the site is Stauffer Chemical Co. in Tarpon Springs (EPA ID FLD010596013). For more information on any of those sites, see EPA, Superfund information systems, CERCLIS Database, www.epa.gov/superfund/sites/cursites/index.htm.

64. B.F. McPherson and R. Halley, "The South Florida environment: a region under stress," USGS Circular 1134, sofia.usgs.gov/publications/circular/1134/wes/chw.html; and "Groups threaten selenium lawsuit," Idaho Falls *Post Register* Sept. 11, 2003:B1.

65. Pacific Environmental Services, Inc., *Background Report AP-42 Section 6.10 Phosphate Fertilizers*, report prepared for EPA (Research Triangle Park, NC, 1996), pp. 2–3; and World Bank, *Pollution Prevention and Abatement Handbook* (Washington, DC, 1998), p. 387.

66. Kongshaug, *Energy Consumption*.

67. Pacific Environmental Services, *Phosphate Fertilizers*, pp. 10–12.

68. World Bank, *Pollution Prevention*, p. 387.

69. K. Kurt and M. Nelson, "Oklahoma accuses Arkansas poultry companies of polluting its water," Associated Press, July 21, 2005.

70. NSTC, *Hypoxia*, p. 9; Rabalais, Turner, and Scavia, "Beyond science"; and N.N. Rabalais, executive director, Louisiana Universities Marine Consortium, email to Center for Science in the Public Interest, June 14, 2002.

71. NSTC, *Hypoxia*, p. 3; and Rabalais, email.

72. "Hypoxia, the Gulf of Mexico's summertime foe," *Watermarks* Sept. 2004(26):3–5; www.lacoast.gov/watermarks/2004-09/watermarks-2004-10.pdf.

73. J.R. Dandelski, *Marine Dead Zones: Understanding the Problem*, Congressional Research Service Report for Congress (1998), www.ncseonline.org/nle/crsreports/marine/mar-30.cfm.

74. NSTC, *Hypoxia*, pp. 4–5.

75. "Link between agricultural runoff and massive algal blooms in the sea," *Medical News Today* Dec. 8, 2004, www.medicalnewstoday.com/medicalnews.php?newsid=17524; and J.M. Beman, K.R. Arrigo, and P.A. Matson, "Agricultural runoff fuels large phytoplankton blooms in vulnerable areas of the ocean," *Nature* (2005) 434:211–14.

76. EPA, "Pretreatment program," http://cfpub.epa.gov/npdes/home.cfm?program_id=3, accessed Dec. 28, 2005.

77. "U.S. sets new farm-animal pollution curbs," *New York Times* Dec. 17, 2002:D28.

78. EPA, *Draft Guidance Manual and Example NPDES Permit for Concentrated Animal Feeding Operations* (1999), www.epa.gov/npdes/pubs/dman_afo.pdf, p. 8.

79. Illinois Environmental Protection Agency, *Understanding the Pollution Potential of Livestock Waste* (Springfield, 1991).

80. Pew Oceans Commission, *Marine Pollution in the United States* (Arlington, VA: Pew Charitable Trusts, 2001), p. 29.

81. National Research Council, *Air Emissions from Animal Feeding Operations: Current Knowledge, Future Needs* (Washington, DC: National Academies Press, 2003), p. 52.

82. M.A. Mallin, J.M. Burkholder, and L.B. Cahoon, "The North and South Carolina coasts," *Marine Poll Bull* (2000) 41:56–75.

83. R. Kellogg, R. Nehring, A. Grube, et al., "Trends in the potential for environmental risk from pesticide loss from farm fields" (USDA NRCS, 1999), www.nrcs.usda.gov/technical/land/pubs/pesttrend.html.

84. EPA, *Pesticides in Drinking-Water Wells*, Pub. 20T-1004 (1990), www.pueblo.gsa.gov/cic_text/housing/water-well/waterwel.txt.

85. G. Wolff, *Investing in Clean Agriculture: How California Can Strengthen Agriculture, Reduce Pollution and Save Money* (Oakland, CA: Pacific Institute for Studies in Development, Environment, and Security, 2005), p. 12.

86. D.W. Kolpin, J.E. Barbash, and R.J. Gilliom, "Occurrence of pesticides in shallow groundwater of the United States: initial results from the National Water-Quality

Assessment Program," *Environ Sci Technol* (1998) 32:558–66. Similar results were found in a newer USGS study, *Pesticides in the Nation's Streams and Ground Water, 1992–2001—A Summary*, http://pubs.usgs.gov/fs/2006/3028/pdf/fs2006-3028.pdf.

87. USGS, *Herbicides in Rainfall across the Midwestern and Northeastern United States, 1990–91* (1998), http://ks.water.usgs.gov/Kansas/pubs/fact-sheets/fs.181-97.html.

88. USGS, "Glyphosate herbicide found in many Midwestern streams, antibiotics not common," http://toxics.usgs.gov/highlights/glyphosate02.html, accessed Mar. 19, 2004.

89. D.W. Kolpin, E.T. Furlong, M.T. Meyer, et al., "Pharmaceuticals, hormones, and other organic wastewater contaminants in U.S. streams, 1999–2000: a national reconnaissance," *Environ Sci Technol* (2002) 36:1202–11.

Argument #5. Cleaner Air (pp. 103-112)

1. Associated Press, "Jury selection begins in case against dairy farmer for 2 deaths," Sept. 10, 2004, www.sfgate.com/cgi-bin/article.cgi?f=/news/archive/2004/09/10/state1644 EDT0112.DTL.

2. U.S. Department of Agriculture, National Agricultural Statistics Service (USDA NASS), Quick Stats: Agricultural Statistics Data Base, www.nass.usda.gov:8080/QuickStats/, accessed June 4, 2006.

3. Meat production: USDA NASS, *Farm Numbers*; chicken production: USDA, Economic Research Service, *Data Product: Poultry Yearbook* (2004), www.ers.usda.gov/data/sdp/view.asp?f=livestock/89007/.

4. Potash and Phosphate Institute and Potash and Phosphate Institute of Canada (PPI-PPIC), *Technical Bulletin 2002–1: Plant Nutrient Use in North American Agriculture—Producing Food and Fiber, Preserving the Environment, Integrating Organic and Inorganic Sources* (Norcross, GA, 2002), p. 58.

5. R. Koelsch, *Environmental Considerations for Manure Application System Selection*, G95-1266-A (University of Nebraska-Lincoln Extension Publications, 1996), http://ianrpubs.unl.edu/wastemgt/g1266.htm.

6. PPI-PPIC, *Technical Bulletin*, p. 58.

7. J. Barker, *Safety in Swine Production Systems* (Raleigh: North Carolina Cooperative Extension Agency, 1996), www.bae.ncsu.edu/programs/extension/publicat/wqwm/pih104.html.

8. P. Viney, V.P. Aneja, J.P. Chauhan, and J.T. Walker, "Characterization of atmospheric ammonia emissions from swine waste storage and manure lagoons," *J Geophys Res-Atmos* (2000) 105:11,535–45.

9. J.A. Zahn, A. Tung, B. Roberts, et al., "Abatement of ammonia and hydrogen sulfide emissions from a swine lagoon using a polymer biocover," *J Air Waste Manag Asso* (2001) 51:562–73.

10. C.E. Pitcairn, U.M. Skiba, M.A. Sutton, et al., "Defining the spatial impacts of poultry farm ammonia emissions on species composition of adjacent woodland groundflora using Ellenberg Nitrogen Index, nitrous oxide and nitric oxide emissions and foliar nitrogen as marker variables," *Environ Pollut* (2002) 119:9–21.

11. National Research Council (NRC), *Air Emissions from Animal Feeding Operations: Current Knowledge, Future Needs* (Washington, DC: National Academies Press, 2003), p. 52.

12. NRC, *Air Emissions*, p. 72.

13. Ontario Medical Association, *Ground Level Ozone Position Paper*, www.oma.org/health/smog/ground.asp.

14. T. Pelton, "Critics charge animal farms are feeding pollution into air," *Baltimore Sun* Feb. 2, 2005:1A.

15. Environmental Law & Policy Center, "Illinois rivers protection initiative," www.elpc. org/forest/water/ammonia.htm.

16. Pelton, "Animal farms."

17. A. Martin, "Livestock industry finds friends in EPA: document details lobbyists' impact on air-quality plan," *Chicago Tribune* May 16, 2004:C9.

18. U.S. Environmental Protection Agency (EPA), *Inventory of U.S. Greenhouse Gas Emissions and Sinks: 1990–1998*, EPA 236–R-00–001 (2000), http://yosemite.epa.gov/oar/ globalwarming.nsf/UniqueKeyLookup/SHSU5BMQ76/$File/2000-inventory.pdf.

19. B. Field, *Beware On-Farm Manure Storage Hazards* (West Lafayette, IN: Purdue University Cooperative Extension Service, 1980), www.agcom.purdue.edu/AgCom/Pubs/S/S-82.html; and *Preventing Deaths of Farm Workers in Manure Pits*, NIOSH Publication 90-103, (National Institute for Occupational Safety and Health, 1990), www.cdc.gov/niosh/90-103.html.

20. NRC, *Air Emissions*, p. 54.

21. EPA, *Inventory*. Methane emissions from livestock and manure total 54.8 million metric tons of carbon equivalent (multiply by 3.67 to convert to carbon dioxide). The EPA estimates that the average automobile emits 6.14 metric tons of carbon dioxide per year. EPA, "Personal greenhouse gas calculator," http://yosemite.epa.gov/OAR/ globalwarming.nsf/content/ResourceCenterToolsGHGCalculator.html.

22. PPI-PPIC, *Technical Bulletin*, pp. 60–61; and NRC, *Air Emissions*, p. 52.

23. NRC, *Air Emissions*, p. 52.

24. U.S. Department of Energy (DOE), "Nitrous oxide emissions" (2001), www.eia.doe.gov/ oiaf/1605/gg00rpt/nitrous.html#nap, accessed Aug. 5, 2004.

25. V. Smil, *Cycles of Life* (New York: Scientific American, 1997), p. 136.

26. NRC, *Air Emissions*, p. 52.

27. NRC, *Air Emissions*, pp. 51, 53.

28. NRC, *Air Emissions*, p. 21.

29. Pacific Environmental Services, Inc., *Background Report AP-42 Section 6.8 Ammonium Nitrate Fertilizer*, report prepared for EPA (Research Triangle Park, NC, 1996), p. 5; and EPA. *AP-42: Compilation of Air Pollution Emission Factors, Vol. 1.*, 5th ed. (Washington DC, 1995), p. 8.3–3.

30. EPA, *AP-42*, pp. 8.1-4, 8.8-4, 8.3-7; and EPA, "Effects of acid rain: human health" (2003), www.epa.gov/airmarkt/acidrain/effects/health.html, accessed Apr. 3, 2005.

31. G. Kongshaug, *Energy Consumption and Greenhouse Gas Emissions in Fertilizer Production* (International Fertilizer Industry Association, 1998), www.fertilizer. org/ifa/publicat/PDF/1998_biblio_65.pdf.

32. Adapted from Kongshaug, *Energy Consumption*; K.J. Hulsbergen and W.D. Kalk, "Energy balances in different agricultural systems: can they be improved?," paper presented at International Fertiliser Society Symposium, Lisbon, Mar. 5, 2001 (York, UK: International Fertiliser Society, 2001), p. 8; and DOE, "Energy consumption estimates by source, 1960–2000, United States" (2003), www.eia.doe. gov/emeu/states/sep_use/total/use_tot_us.html.

33. B.J. Nebel, *Environmental Science: The Way the World Works*, 3rd ed. (Englewood Cliffs, NJ: Prentice Hall, 1990), pp. 300–09.

34. NRC, *Air Emissions*, pp. 69–71.

35. Centers for Disease Control and Prevention (CDC), National Agricultural Safety Database: Manure Gas, Hydrogen Sulfide (2002), www.cdc.gov/nasd/docs/d001501– d001600/d001535/d001535.html, accessed Mar. 12, 2003.

36. Barker, *Safety*.

37. Barker, *Safety*.

38. J. Lee, "Neighbors of vast hog farms say foul air endangers their health," *New York Times* May 11, 2003:1.

39. Barker, *Safety*.

40. CDC, database.

41. NRC, *Air Emissions*, p. 54.

42. NRC, *Air Emissions*, pp. 68–69; and S.S. Schiffman, "Livestock odors: implications for human well-being," *J Anim Sci* (1998) 76:1343–55.

43. Schiffman, "Livestock odors"; and NRC, *Air Emissions*, p. 68.

44. C.M. Williams, *Benefits of Adopting Environmentally Superior Swine Waste Management Technologies in North Carolina: An Environmental and Economic Assessment* (Research Triangle Park, NC: RTI International, 2003), pp. 6.1–6.3.

45. NRC, *Air Emissions*, p. 56.

46. Schiffman, "Livestock odors"; NRC, *Air Emissions*, pp. 68–69; and R.C. Avery, S. Wing, S.W. Marshall, et al., "Odor from industrial hog farming operations and mucosal immune function in neighbors," *Arch Environ Health* (2004) 59(2):101–08.

47. Schiffman, "Livestock odors"; and NRC, *Air Emissions*, p. 68.

48. NRC, *Air Emissions*, p. 55.

49. A.R. Chapin, A. Rule, K. Gibson, et al., "Airborne multi-drug resistant bacteria isolated from a concentrated swine feeding operation," *Environ Health Perspect* (2005) 113(2), http://ehp.niehs.nih.gov/members/2004/7473/7473.pdf.

50. G. Hamscher, H.T. Pawelzick, S. Sczesny, et al., "Antibiotics in dust originating from a pig-fattening farm: a new source of health hazard for farmers?," *Environ Health Perspect* (2003) 111:1590–94.

51. USDA, Agricultural Research Service, "Action plan: Component V: pesticides and other synthetic organic compounds," www.ars.usda.gov/research/programs/programs.htm?np_code=203&docid=324.

52. U.S. Geological Survey, *Herbicides in Rainfall across the Midwestern and Northeastern United States, 1990–91* (1998), http://ks.water.usgs.gov/Kansas/pubs/fact-sheets/fs.181-97.html.

53. NRC, *Air Emissions*, p. 55.

54. J. Eilperin, "In California, agriculture takes center stage in pollution debate," *Washington Post* Sept. 26, 2005:A1.

Argument #6. Less Animal Suffering (pp. 113-139)

1. *Congressional Record* July 9, 2001:S7310–11.

2. U.S. Department of Agriculture, National Agricultural Statistics Service (USDA NASS), *Livestock Slaughter 2002 Summary* (Washington, DC, 2003), p. 1.

3. USDA NASS, *Poultry Slaughter 2002 Summary* (Washington, DC, 2003), pp. 2–3.

4. American Meat Institute, Animal handling frequently asked questions, www.animalhandling.org/faqs.htm.

5. D. Barboza, "Animals seeking happiness," *New York Times* June 29, 2003:4-5.

6. V. Hirsch, *Legal Protections of the Domestic Chicken in the United States and Europe* (Michigan State University, Detroit College of Law, Animal Legal and Historical Center, 2003), www.animallaw.info/articles/dduschick.htm#3.

7. For cattle, hogs, and sheep slaughtered in commercial plants and on farms: USDA NASS, *Livestock Slaughter 2004 Summary,* http://usda.mannlib.cornell.edu/reports/nassr/livestock/pls-bban/lsan0305.pdf; for poultry: USDA NASS, *Poultry Slaughter 2004 Annual Summary,* http://usda.mannlib.cornell.edu/reports/nassr/poultry/ppy-bban/pslaan05.pdf. These numbers omit hundreds of millions of additional animals (mostly chickens) that die (due to injury or illness) or are killed (such as male chicks by the egg industry) before they got to slaughterhouses.

8. M. Scully, *Dominion: The Power of Man, the Suffering of Animals, and the Call to Mercy* (New York: St. Martin's Griffon, 2002).

9. B.E. Rollin, *Farm Animal Welfare* (Ames, IA: Iowa State University Press, 1995), p. 100.

10. C.W. Arave and J.L. Albright, "Animal welfare issues: dairy," in R.D. Reynells and B.R. Eastwood, eds., *Animal Welfare Issues Compendium: A Collection of 14 Discussion Papers* (Washington, DC: USDA, Cooperative State Research Extension Education Service, Plant and Animal Production, Protection and Processing, 1997), www.nal.usda.gov/awic/pubs/97issues.htm, p. 63; Rollin, *Farm Animal Welfare,* pp. 102–03; and C. Phillips, *Cattle Behaviour and Welfare,* 2nd ed. (Oxford: Blackwell Publishing, 2002), p. 211.

11. S.M. Abutarbush and O.M. Radostits, "Obstruction of the small intestine caused by a hairball in 2 young calves," *Can Vet J* (2004) 45(4):324–25.

12. Phillips, *Cattle Behaviour.*

13. T. Field, "Effects of hot iron branding on value of cattle hides," *The Final Report of the National Beef Quality Audit, 1991* (Englewood, CO: National Cattlemen's Association, 1992), p. 127; cited in Rollin, *Farm Animal Welfare,* p. 58.

14. Arave and Albright, "Dairy," p. 60.

15. Rollin, *Farm Animal Welfare,* p. 61; and Rollin, university distinguished profess or, Colorado State University, email to Center for Science in the Public Interest (CSPI), Aug 24, 2004.

16. Arave and Albright, "Dairy," p. 61.

17. Rollin, *Farm Animal Welfare,* p. 62.

18. Rollin, *Farm Animal Welfare,* pp. 62–63.

19. R. Cobb, "Horns on domestic farm animals," Working with Farm Animals course materials, University of Illinois at Urbana-Champaign, http://classes.aces.uiuc.edu/AnSci103/horns.html.

20. Arave and Albright, "Dairy," p. 61.

21. Rollin, *Farm Animal Welfare,* p. 105.

22. Phillips, *Cattle Behaviour,* p. 214.

23. Ministry of Agriculture, Food and Fisheries, *The Animal Welfare Act/The Animal Welfare Ordinance* (2004), www.sweden.gov.se/content/1/c6/01/89/74/356685f8.pdf; and R. Silvanic, "Dairy production in Sweden," www.vetmed.iastate.edu/academics/international/recenttrips/sweden2003/studentpapers/swedendairySilvanic.pdf.

24. Rollin, *Farm Animal Welfare,* p. 99.

25. Rollin, *Farm Animal Welfare,* p. 119.

26. D.E. Granstrom, "Agricultural (nonbiomedical) animal research outside the laboratory: a review of guidelines for institutional animal care and use committees," *ILAR J* (2003) 44(3): 206–10.

27. J.A. Mench and P.B. Siegel, "Animal welfare issues: poultry," in Reynells and Eastwood, eds., *Animal Welfare Issues Compendium,* p. 105; and Rollin, *Farm Animal Welfare,* p. 134.

28. M.E. Ensminger, *Animal Science,* 9th ed. (Danville, IL: Interstate Publishing, 1991), p. 184.

29. The Mini Cooper is 142.8 by 75.8 inches, or 75.2 square feet. "Mini Features and Specs, 2003," BMW of North America, www.miniusa.com/link/ourcars/features/minicooper/exterior/dimensions/none.

30. S.L. Davis and P.R. Cheek, "Do domestic animals have minds and the ability to think? A provisional sample of opinions on the question," *J Anim Sci* (1998) 76:2072–79.

31. S.A. Ewing, D.C. Lay, E. von Berell, *Farm Animal Well-Being: Stress Physiology, Animal Behavior, and Environmental Design* (Upper Saddle River, NJ: Prentice Hall, 1999), p. 222; Rollin, *Farm Animal Welfare*, pp. 76, 91; and Alberta Pork, "What is a gestation crate?," www.albertapork.com/news.aspx?NavigationID=1456.

32. Rollin, *Farm Animal Welfare*, p. 93.

33. Food and Agriculture Organization of the United Nations (FAO), FAOSTAT, http://apps.fao.org/faostat/collections?version=ext&hasbulk=0&subset=agriculture, accessed Aug. 11, 2004; and Department for Environment and Rural Affairs, "Introduction to veterinary surveillance and emerging diseases," in *Animal Health 2000; The Chief Veterinary Officer's Report for 2000* (London, 2001), ch. A4.

34. J. Barker, *Safety in Swine Production Systems* (Raleigh: North Carolina Cooperative Extension Agency, 1996), www.bae.ncsu.edu/programs/extension/publicat/wqwm/pih104.html.

35. Compassion Over Killing, "About ISE," www.isecruelty.com/aboutise.php.

36. Ewing, Lay, and von Berell, *Farm Animal Well-Being*, p. 250.

37. United Egg Producers, *United Egg Producers Animal Husbandry Guidelines for U.S. Egg Laying Flocks, 2005*, 2nd ed., www.uepcertified.com/docs/2005_UEPanimal_welfare_guidelines.pdf.

38. USDA, "USDA releases estimates of farm production losses," Release No. 0385.05, Sept. 20, 2005.

39. Scully, *Dominion*.

40. Mench and Siegel, "Poultry," p. 101; and "Laying down minimum standards for the protection of laying hens," *Official Journal of the European Communities*, Council Directive 1999/74/Ec, http://europa.eu.int/eur-lex/pri/en/oj/dat/1999/l_203/l_20319990803en00530057.pdf.

41. Rollin, *Farm Animal Welfare*, p. 119.

42. Rollin, *Farm Animal Welfare*, pp. 120–26.

43. C. Druce and P. Lymbery, *Outlawed in Europe: Three Decades of Progress in Europe* (Animal Rights International, 2001), www.ari-online.org/pages/europe1.html.

44. S. Romero, "Virus takes a toll on Texas poultry industry," *New York Times* May 16, 2003: C1; and "Avian flu found on Maryland farm," *Washington Post* Mar. 7, 2004:C3.

45. Rollin, *Farm Animal Welfare*, p. 133.

46. Arave and Albright, "Dairy," p. 64.

47. B. Faye, F. Lescourret, N. Dorr, et al., "Interrelationships between herd management practices and udder health status using canonical correspondence analysis," *Prev Vet Med* (1997) 32:171–92.

48. Arave and Albright, "Dairy."

49. Rollin, *Farm Animal Welfare*, p. 106; and Arave and Albright, "Dairy," p. 59.

50. I.R. Dohoo, K. Leslie, L. DesCôteaux, et al., "A meta-analysis review of the effects of recombinant bovine somatotropin: 1. Methodology and effects on production, 2. Effects on animal health, reproductive performance, and culling," *Can J Vet Res* (2003) 67(4):241-64; and Monsanto, "Posilac," www.monsantodairy.com/.

51. Rollin, *Farm Animal Welfare*, p. 125.

52. United Egg Producers, *Animal Husbandry Guidelines.*

53. Rollin, *Farm Animal Welfare*, pp. 103–04; Ewing, Lay, and von Berrell, *Farm Animal Well-Being*, pp. 189–91; and Phillips, *Cattle Behavior*, p. 210.

54. Ewing, Lay, and von Berrell, *Farm Animal Well-Being*, pp. 189–91.

55. Ewing, Lay, and von Berrell, *Farm Animal Well-Being*, pp. 189–91.

56. Phillips, *Cattle Behavior*, p. 213.

57. Ewing, Lay, and von Berrell, *Farm Animal Well-Being*, p. 220.

58. Y. Hyun, M. Ellis, G. Riskowski, and R.W. Johnson, "Growth performance of pigs subjected to multiple concurrent environmental stressors," *J Anim Sci* (1998) 76:721–77.

59. Ewing, Lay, and von Berrell, *Farm Animal Well-Being*, p. 186.

60. P.J. Holden and J. McGlone, "Animal welfare issues: swine," in Reynells and Eastwood, *Animal Welfare Issues Compendium*, p. 127.

61. Rollin, *Farm Animal Welfare*, p. 75.

62. Ewing, Lay, and von Berrell, *Farm Animal Well-Being*, p. 220.

63. Ewing, Lay, and von Berrell, *Farm Animal Well-Being*, pp. 179 and 194–95.

64. Rollin, *Farm Animal Welfare*, p. 119.

65. Rollin, *Farm Animal Welfare*, p. 121.

66. Rollin, *Farm Animal Welfare*, p. 122.

67. A.B. Webster, "Behavior of chickens" in D.D. Bell and W.D. Weaver, eds., *Commercial Chicken Meat and Egg Production* (Norwell, MA: Kluwer Academic Publishers, 2002), pp. 71–86.

68. Ewing, Lay, and von Berrell, *Farm Animal Well-Being*, p. 194.

69. Rollin, *Farm Animal Welfare*, p. 133.

70. Compassion Over Killing, *A COK Report: Animal Suffering in the Broiler Industry* (Washington, DC, 2004).

71. C.J. Savory, K. Maros, and S.M. Rutter, "Assessment of hunger in growing broiler breeders in relation to a commercial restricted feeding programme," *Animal Welfare* (1993) 2:131–52; and C.J. Savory and K. Maros, "Influence of degree of food restriction, age, and time of day on behaviour of broiler breeder chickens," *Behavioural Processes* (1993) 29:179–90.

72. D. Sainsbury, *Animal Health*, 2nd ed. (Malden, MA: Blackwell Science Ltd, 1998), p. 2.

73. Mench and Siegel, "Poultry," p. 101.

74. Sainsbury, *Animal Health*, p. 2.

75. M.E. Ensminger and R.C. Perry, *Beef Cattle Science*, 7th ed. (Danville, IL: Interstate Publishing, 1997), pp. 300–06; E. Schlosser, *Fast Food Nation* (New York: HarperCollins Perennial, 2002), p. 202; R.D. Shaver, "By-product feedstuffs in dairy cattle diets in the Upper Midwest," www.wisc.edu/dysci/uwex/nutritn/pubs/ByProducts/ByproductFeed stuffs.html; and S.B. Blezinger, "Energy issues affect choices for cattle feed ingredients," *Cattle Today Online*, www.cattletoday.com/archive/2005/October/CT421.shtml.

76. USDA Economic Research Service (USDA ERS), www.ers.usda.gov/Data/FoodConsumption/FoodAvailQueriable.aspx, accessed Aug. 11, 2005.

77. Rollin, *Farm Animal Welfare*, pp. 111–13.

78. R.H. Poppenga, "Current environmental threats to animal health and productivity," *Vet Clin North Am Food Anim Pract* (2000) 16:545–58.

79. U.S. Food and Drug Administration (FDA), *Food and Drug Administration Pesticide Program: Residue Monitoring 2000* (Washington, DC, 2001), p. 12.

80. S.M. Rhind, "Endocrine disrupting compounds and farm animals: their properties, actions and routes of exposure," *Domest Anim Endocrinol* (2002) 23:179–87.

81. V. Ishler, J. Heinrichs, and G. Varga, *From Feed to Milk: Understanding Rumen Function,"* Penn State Extension Circular 422 (1996), www.das.psu.edu/dairynutrition/documents/rumen.pdf, p. 10; J.C. Plazier, "Feeding forage to prevent rumen acidosis in cattle" (University of Manitoba, 2002), www.umanitoba.ca/afs/fiw/020704.html; and J. Couzin, "Cattle diet linked to bacterial growth," *Science* (1998) 281:1578.

82. F. Diez-Gonzalez, T.R. Callaway, M.G. Kizoulis, et al., "Grain feeding and the dissemination of acid-resistant *Escherichia coli* from cattle," *Science* (1998) 281:1666–68; and J.B. Russell, F. Diez-Gonzalez, and G.N. Jarvis, "Potential effect of cattle diets on the transmission of pathogenic *Escherichia coli* to humans," *Microbes Infect* (2000) 2:45–53.

83. "High-grain cattle diets cause drug need," *Meat Processing* May 23, 2001, www.meatnews.com/index.cfm?fuseaction=Article&artNum=1157.

84. D. Griffin, L. Perino, and D. Hudson, *Feedlot Lameness* (Lincoln, NE: University of Nebraska, 1993), p. 1.

85. "Grain overload," *Merck Veterinary Manual*, 9th ed. (2005), www.merckvetmanual.com/mvm/index.jsp?cfile=htm/bc/21703.htm&word=high%2cgrain%2cdiet.

86. "Cattle die after feedlot seized," *Toronto Star* Jan. 10, 2005:A4; and "Grain overload."

87. "High-grain cattle diets cause drug need."

88. Texas Cooperative Extension, "Animal disorders: bloat," http://stephenville.tamu.edu/~butler/foragesoftexas/animaldisorders/bloat.html.

89. "Cattle die after feedlot seized."

90. Ewing, Lay, and von Berrell, *Farm Animal Well-Being*, pp. 189–91.

91. Phillips, *Cattle Behavior*, pp. 210–11.

92. Z.O. Müller, "Economic aspects of recycled wastes," in *New Feed Resources: Proceedings of a Technical Consultation Held in Rome, 22–24 Nov. 1976* (FAO), www.fao.org/DOCREP/004/X6503E/X6503E14.htm.

93. Plazier, "Feeding forage"; and J.B. Russell, F. Diez-Gonzalez, and G.N. Jarvis, "Effects of diet shifts on *Escherichia coli* in cattle," *J Dairy Sci* (2000) 83(4):869.

94. The Innovation Group, "Sodium bicarbonate," profile, www.the-innovation-group.com/ChemProfiles/Sodium%20Bicarbonate.htm.

95. Rhind, "Endocrine disrupting compounds."

96. H.B. Sewell, *Growth Stimulants (Implants)*, University of Missouri-Columbia Agricultural Pub. G2090 (1993), http://extension.missouri.edu/explore/agguides/ansci/g02090.htm.

97. European Commission, *Opinion of the Scientific Committee on Veterinary Measures Relating to Public Health: Assessment of Potential Risks to Human Health from Hormone Residues in Bovine Meat and Meat Products* (1999), http://europa.eu.int/comm/food/fs/sc/scv/out21_en.pdf.

98. FDA, Center for Veterinary Medicine, "The use of steroid hormones for growth promotion in food-producing animals" (2002), www.fda.gov/cvm/hormones.htm; USDA Foreign Agriculture Service, "A primer on beef hormones" (1999), http://www.useu.be/issues/BeefPrimer022699.html; and World Health Organization, *Evaluation of Certain Veterinary Drug Residues in Food: 52nd Report of the Joint FAO/WHO Expert Committee on Food Additives* (2000), http://whqlibdoc.who.int/trs/WHO_TRS_893.pdf.

99. Confidential email to CSPI, May 30, 2006.

100. J. Raloff, "Hormones: here's the beef," *Science News* (2002) 161:10; E.F. Orlando, A.S. Kolok, G. Binzcik, et al., "Endocrine-disrupting effects of cattle feedlot effluent on an aquatic sentinel species, the fathead minnow," *Environ Health Perspect* (2004) 112(5):A270; and U.S. Environmental Protection Agency, "Funding opportunities: fate and effects of hormones in waste from concentrated animal feeding operations (CAFOS)," http://es.epa.gov/ncer/rfa/2006/2006_star_cafos.html.

101. E.F. Orlando, reproductive biologist, Florida Atlantic University, email to CSPI, May 16, 2006.

102. E. Weise, "Iowa, Minnesota are latest to test for dioxin in animal-feed probe," *USA Today* Mar. 26, 2003:9D.

103. J. Lee, "Sewer sludge spread on fields is fodder for lawsuits," *New York Times* June 26, 2003:A20.

104. R.L. Mahler, P. Ernestine, and R. Taylor, *Nitrate and Groundwater* (Moscow, ID: University of Idaho, 2002), www.uidaho.edu/wq/wqpubs/cis872.html.

105. D.G. McNeil Jr., "KFC supplier accused of cruelty to animals," *New York Times* July 20, 2004:C2; and Pub. L. No. 95-445, 92 Stat. 1069 (1978).

106. As cited in Poppenga, "Current environmental threats."

107. D. Grady and D.G. McNeil Jr, "Rules issued on animal feed and use of disabled cattle," *New York Times* Jan. 27, 2004:A12.

108. D.A. Shields and K.A. Mathews, *Interstate Livestock Species* (Washington, DC: USDA ERS, 2003), p. 4.

109. Ewing, Lay, and von Berrell, *Farm Animal Well-Being*, p. 241.

110. Shields and Mathews, *Interstate Livestock*, p. 4.

111. N.G. Gregory, *Animal Welfare and Meat Science* (New York: CABI Publishing, 1998), p. 18.

112. T. Grandin, "Perspectives on transportation issues: the importance of having physically fit cattle and pigs" (2000), www.grandin.com/behaviour/perspectives.transportation.issues.html.

113. S.D. Eischer, "Transportation of cattle in the dairy industry: current research and future directions," *J Dairy Sci* (2001) 84(suppl.):E19–23.

114. Rollin, *Farm Animal Welfare*, p. 106.

115. J.F. Currin and W.D. Whittier, *Feeder and Stocker Health and Management Practices* (Blacksburg, VA: Virginia Cooperative Extension, 2000), p. 1.

116. N.R. Hartwig, "Bovine respiratory disease" (Iowa Beef Industry Council), www.iabeef.org/Content/brd.aspx.

117. Gregory, *Animal Welfare*, p. 35; and D.G. McNeil Jr., "Inquiry finds lax federal inspections at kosher meat plant," *New York Times* Mar. 10, 2006:A13.

118. Rollin, *Farm Animal Welfare*, p. 135.

119. L. Compa, *Blood, Sweat, and Fear: Workers' Rights in U.S. Meat and Poultry Plants* (New York: Human Rights Watch, 2004), p. 34.

120. World Society for the Protection of Animals, *Industrial Animal Agriculture: The Next Global Health Crisis?* (London, 2004), p. 10.

121. Compa, *Blood, Sweat, and Fear*, p. 40.

122. J. Motavalli, "The case against meat," *E/Environ Mag* (2002) 13(1):5.

123. Compa, *Blood, Sweat, and Fear*, pp. 33, 38–40, 42–43.

124. S. Greenhouse, "Rights group condemns meatpackers on job safety," *New York Times* Jan. 26, 2005:A13.

125. Schlosser, *Fast Food Nation*, p. 178.

126. T. Grandin, *Survey of Federally Inspected Beef, Veal, Pork, and Sheep Slaughter Plants* (Washington, DC: USDA Agricultural Research Service, 1997).

127. Gregory, *Animal Welfare*, p. 15; and Grandin, *Stunning and Handling*, tables 1–3.

128. Mench and Siegel, "Poultry," p. 104.

129. Humane Farming Association, "HFA's petition to Washington State, affidavit #16" (2005), www.hfa.org/hot_topic/wash_petition2.html.

130. Rollin, *Farm Animal Welfare*, pp. 69–70.

131. M. Warner, "Sharpton joins with an animal activist group in calling for a boycott of KFC," *New York Times* Feb. 2, 2005:C1.

132. Mench and Siegel, "Poultry," p. 104.

133. U.S. Government Accountability Office, *Humane Methods of Slaughter Act: USDA Has Addressed Some Problems but Still Faces Enforcement Challenges* (2004), www.gao.gov/new.items/d04247.pdf, p. 1; Pub. L. No. 95–445, 92 Stat. 1069 (1978); and E. Williamson, "Humane Society to sue over poultry slaughtering," *Washington Post* Nov. 21, 2005:B2.

134. USDA ERS, *Agricultural Resources and Environmental Indicators* (Washington, DC, 2003), p. 3.1-9.

135. USDA ERS, *Agricultural Resources*, p. 3.1-12.

136. National Research Council (NRC), *The Future Role of Pesticides in U.S. Agriculture* (Washington, DC: National Academies Press, 2000), p. 19.

137. See American Beekeeping Federation, 2006 ABF Resolution CR12, pesticide registration process, http://abfnet.org/?page_id=42; and North American Pollinator Protection Campaign, "Plans and projects," www.nappc.org/plansEn.html.

138. NRC, *Future Role of Pesticides*, p. 82.

139. M. Deinlein, *When It Comes to Pesticides, Birds Are Sitting Ducks*, Smithsonian Migratory Bird Center Fact Sheet No. 8, http://nationalzoo.si.edu/ConservationAndScience/MigratoryBirds/Fact_Sheets/fxsht8.pdf.

140. NRC, *Future Role of Pesticides*, p. 80.

Changing Your Own Diet (pp. 143-150)

1. American Cancer Society, "The complete guide—nutrition and physical activity," www.cancer.org/docroot/PED/content/PED_3_2X_Diet_and_Activity_Factors_That_Affect_Risks.asp; American Diabetes Association, "Evidence-based nutrition principles and recommendations for the treatment and prevention of diabetes and related complications, *Diabetes Care* (2002) 25:S50–60; A.H. Lichtenstein, L.J. Appel, M. Brands, et al., "Diet and lifestyle recommendations revision 2006: a scientific statement from the American Heart Association Nutrition Committee," *Circulation* (2006) 114; American Heart Association, "Our 2006 diet and lifestyle recommendations," www.americanheart.org/presenter.jhtml?identifier=851; American Institute for Cancer Research/World Cancer Research Fund, *Food, Nutrition, and the Prevention of Cancer: A Global Perspective* (Washington, DC: American Institute for Cancer Research, 1997); U.S. Department of Health and Human Services and U.S. Department of Agriculture, *Dietary Guidelines for Americans* (2005), www.health.gov/dietaryguidelines/dga2005/document/pdf/DGA2005.pdf; and World Health Organization, "Obesity and overweight" (Geneva, 2003), www.who.int/dietphysicalactivity/publications/facts/obesity/en/.

2. National Heart, Lung, and Blood Institute (NHLBI), *Facts about the DASH Eating Plan* (rev. 2003), www.nhlbi.nih.gov/health/public/heart/hbp/dash/new_dash.pdf.

3. Adapted from NHLBI, *DASH Eating Plan*, p. 5.

4. American Dietetic Association and Dietitians of Canada, "Position of the American Dietetic Association and Dietitians of Canada: vegetarian diets," *J Am Diet Assoc* (2003) 103:748–65; and G.E. Fraser, *Diet, Life Expectancy, and Chronic Disease: Studies of Seventh-day Adventists and Other Vegetarians* (New York: Oxford, 2003).

5. American Dietetic Association and Dietitians of Canada, "Vegetarian diets."

6. V. Messina, V. Melina, and A.R. Mangels, "A new food guide for North American vegetarians," *Can J Diet Prac Res* (2003) 64:82–86, www.dietitians.ca/news/downloads/Vegetarian_Food_Guide_for_NA.pdf.

7. Adapted from Messina, Melina, and Mangels, "New food guide."

8. R. Obeid, J. Geisel, H. Schorr, et al., "The impact of vegetarianism on some hematological parameters," *Eur J Haem* (2002) 69:275–79; C. Lamberg-Allardt, M. Karkkainen, R. Seppanen, et al., "Low serum 25–hydroxyvitamin D concentrations and secondary hyperparathyroidism in middle-aged white strict vegetarians," *Am J Clin Nutr* (1993) 58:684–89; and E.H. Haddad, L.S. Berk, J.D. Kettering, et al., "Dietary intake and biochemical, hematologic, and immune status of vegans compared with non-vegetarians," *Am J Clin Nutr* (1999) 70(suppl):586S–93S.

9. Calculations were made using the Eating Impact Calculator on the Center for Science in the Public Interest's Eating Green web site: www.eatinggreen.org.

Changing Government Policies (pp. 151-168)

1. M. Pollan, *The Omnivore's Dilemma: A Natural History of Four Meals* (East Rutherford, NJ: Penguin Press, 2006).

2. L.H. Baumgard, J.K. Sangster, and D.E. Bauman, "Milk fat synthesis in dairy cows is progressively reduced by increasing supplemental amounts of trans-10, cis-12 conjugated linoleic acid (CLA)," *J Nutr* (2001) 131:1764–69.

3. A.M. Fearon, C.S. Mayne, J.A.M. Beattie, et al., "Effect of level of oil inclusion in the diet of dairy cows at pasture on animal performance and milk composition and properties," *J Sci Food Agric* (2004) 84:497–504.

4. Center for Science in the Public Interest (CSPI), *Anyone's Guess: The Need or Nutrition Labeling at Fast-Food and Other Chain Restaurants* (2003), www.cspinet.org/new/pdf/anyone_s_guess_final_web.pdf.

5. Associated Press, "EPA exempts factory farms from high pollution penalties," Jan. 31, 2006.

6. American Public Health Association, "2003–7 Precautionary moratorium on new concentrated animal feed operations," *Association News*, www.apha.org/legislative/policy/2003/2003-007.pdf.

7. Farm Foundation, *The Future of Animal Agriculture in North America* (Oak Brook, IL, 2006), www.farmfoundation.org/projects/04-32ReportTranslations.htm.

8. U.S. Environmental Protection Agency (EPA), "Funding opportunities: fate and effects of hormones in waste from concentrated animal feeding operations (CAFOS)," http://es.epa.gov/ncer/rfa/2006/2006_star_cafos.html.

9. "The curse of factory farms," *New York Times* Aug. 30, 2002:A18.

10. EPA, *Development Document for the Final Revisions to the National Pollutant Discharge Elimination System Regulation and the Effluent Guidelines for Concentrated Animal Feeding Operations* (2002), accessible at http://cfpub.epa.gov/npdes/afo/cafodocs.cfm, pp. 8-1–11; and Chesapeake Bay Foundation, *Manure's Impact on Rivers, Streams, and the Chesapeake Bay* (Annapolis, 2004), p. 18. A soil scientist with the U.S. Department of Agriculture (USDA) who has studied phytase in hogs says that CAFO producers use whatever feed is provided to them by the feed mill and/or the integrator. An obstacle is that hog feed is pelletized, which can render the phytase enzyme less effective, but innovative technology might solve that problem. D.R. Smith, Ph.D., USDA, Agricultural Research Service, email to CSPI, Sept. 10, 2004.

11. EPA, *Development Document*, pp. 8-1–11; and Chesapeake Bay Foundation, *Manure's Impact*.

12. E. Brzostek, Environmental Quality Incentives Program program specialist, USDA, National Resources Conservation Service, email to CSPI, Dec. 22, 2005.

13. Environmental Working Group, *California Water Subsidies* (2004), www.ewg.org/reports/watersubsidies/.

14. C. Dimitri and L. Oberholtzer, "EU and U.S. organic markets face strong demand under different policies," *Amber Waves* (2006) 4(1):12–19.

15. V. Frances, *Fair Agricultural Chemical Taxes* (Washington, DC: Friends of the Earth, 1999), www.foe.org/res/pubs/pdf/factreport.pdf.

16. Soil and Water Conservation Society, *Sharing the Cost: Creating a Working Land Conservation Trust Fund Through a Tax on Agricultural Inputs?* (Ankeny, IA: Soil and Water Conservation Society, 2003). This analysis notes that two federal programs, the Pittman-Robertson Act and the Dingell-Johnson Act, fund wildlife and fisheries conservation, management, education, and restoration programs through taxes on hunting and fishing equipment. Thus, there are precedents for collecting taxes from certain sectors, distributing funds back to the states, and then ensuring that the sectors that pay the taxes benefit from the programs that are funded.

17. Soil and Water Conservation Society, *Sharing the Cost.*

18. Organisation for Economic Co-operation and Development, *Manure Policy and MINAS: Regulating Nitrogen and Phosphorus Surpluses in Agriculture of the Netherlands* (2005), http://appli1.oecd.org/olis/2004doc.nsf/linkto/com-env-epoc-ctpa-cfa(2004)67-final; and R. Naylor, H. Steinfeld, W. Falcon, et al., "Losing the links between livestock and land," *Science* (2005) Dec. 9:1621–22.

19. Farm subsidies are discussed more fully in the following documents: USDA Economic Research Service, "The 2002 Farm Bill: provisions and economic implications," www.ers.usda.gov/Features/farmbill/; D.E. Ray, speaker, *Agricultural Policy for the 21st Century and the Legacy of the Wallaces*, the John Pesek Colloquium on Sustainable Agriculture, Mar. 3–4, 2004, www.wallacechair.iastate.edu/endeavors/pesekcolloquium/ISU-Pesek-Pkg--04-Bro3.pdf; J.E. Frydenlund, *The Erosion of Freedom to Farm*, Backgrounder 1523 (Washington, DC: Heritage Foundation, 2002), www.heritage.org/Research/Agriculture/BG1523.cfm?renderforprint=1; and Environmental Working Group, Farm subsidy database, www.ewg.org:16080/farm/findings.php, accessed May 6, 2006.

20. The $500 million is the shortfall between what ranchers pay and what the federal government pays for range management. Grazing fees to the U.S. Forest Service and Bureau of Land Management (BLM) raise about $6 million a year. However, in 2000–01, the total direct cost, paid by taxpayers, of range management was $132 million. Indirect costs to both agencies for land management planning, habitat management, forest, rangeland research, and other costs in 2001 were as high as $176 million for the Forest Service and $104 million for BLM. The remainder of the federal subsidy is costs assumed by other agencies. K. Moskowitz and C. Romaniello, *Assessing the Full Cost of the Federal Grazing Program* (Tucson: Center for Biological Diversity, 2002), www.biologicaldiversity.org/swcbd/Programs/grazing/Assessing_the_full_cost.pdf.

21. "Laying down minimum standards for the protection of laying hens," *Official Journal of the European Communities*, Council Directive 1999/74/Ec, http://europa.eu.int/eur-lex/pri/en/oj/dat/1999/l_203/l_20319990803en00530057.pdf; and FARM, "Farmed animal treatment," fact sheet, www.wfad.org/about/treatment.htm.

22. M. Scully, *Dominion: The Power of Man, the Suffering of Animals, and the Call to Mercy* (New York: St. Martin's Griffon, 2002).

23. Farm Animal Welfare Council, www.fawc.org.uk/freedoms.htm.

Appendix A. A Bestiary of Foodborne Pathogens (pp. 171-176)

1. Centers for Disease Control and Prevention (CDC), "*Campylobacter* infections: technical information," www.cdc.gov/ncidod/dbmd/ diseaseinfo/campylobacter_t.htm, accessed Oct. 1, 2003.

2. U.S. Food and Drug Administration, Center for Food Safety and Applied Nutrition (FDA CFSAN), *Bad Bug Book: Campylobacter jejuni* (1992), www.cfsan.fda.gov/~mow/chap4.html.

3. FDA CFSAN, *Bad Bug Book: Campylobacter jejuni*; Guillain-Barré Syndrome Foundation International, *GBS: An Overview* (Wynnewood, PA, 2002), www.guillain-barre.com/ overview.html; and J.C. Buzby, T. Roberts, and B. Allos, *Estimated Annual Costs of* Campylobacter-*Associated Guillain-Barré Syndrome*, Agricultural Economics Report No. 756 (Washington, DC: U.S. Department of Agriculture, Economic Research Service, 1997).

4. I.V. Wesley, S.J. Wells, K.M. Harmon, et al., "Fecal shedding of *Campylobacter* and *Arcobacter* spp. in dairy cattle," *Appl Environ Microbiol* (2000) 66(5):1994–2000.

5. CDC, "*Campylobacter*"; and A. Hingley, *Campylobacter: Low Profile Bug Is Food Poisoning Leader* (Washington, DC: FDA, 1999), www.fda.gov/fdac/features/1999/599_bug.html.

6. FDA, "Enroflaxin for poultry; opportunity for hearing," Docket No. 00N-1571, *Fed Reg* (2000) 65(211):64954–65, www.fda.gov/OHRMS/DOCKETS/98fr/103100a.htm.

7. D. Vugia, A. Cronquist, J. Hadler, et al., "Preliminary FoodNet data on the incidence of infection with pathogens transmitted commonly through food—10 states, United States, 2005," *MMWR Weekly* (2006) 55(14):392–95.

8. FDA CFSAN, *Bad Bug Book: Clostridium perfringens*, www.cfsan.fda.gov/~mow/chap11. html.

9. P.S. Mead, L. Slutsker, V. Dietz, et al., "Food-related illness and death in the United States," *Emerging Infectious Diseases* (1999) 5:607–25; and U.S. Department of Agriculture, Economic Research Service (USDA ERS), *Economics of Food-borne Disease* (Washington, DC: Government Printing Office, 2003).

10. B. Van Voris, "Jack in the Box ends *E. coli* suits," *National Law Journal* Nov. 17, 1997.

11. USDA, Food Safety and Inspection Service (USDA FSIS), "Beef … from farm to table," meat preparation fact sheet (2003), www.fsis.usda.gov/Fact_Sheets/Beef_from_Farm_ to_Table/index.asp.

12. A.V. Tutenel, D. Pierard, J. Van Hoof, et al., "Molecular characterization of *Escherichia coli* O157 contamination routes in a cattle slaughterhouse," *J Food Prot* (2003) 66(9):1564– 69; and J.M. McEvoy, A.M. Doherty, J.J. Sheridan, et al., "The prevalence and spread of *Escherichia coli* O157:H7 at a commercial beef abattoir," *J Appl Microbiol* (2003) 95(2):255–66.

13. Vugia et al. "Preliminary FoodNet data."

14. "Meat plants faulted on safety rules," *Washington Post* Feb. 5, 2003:A24.

15. J.A. Crump, A.C. Sulka, A.J. Langer, et al., "An outbreak of *Escherichia coli* O157:H7 infections among visitors to a dairy farm," *N Eng. J Med* (2002) 347(8):555–60; and CDC, "Outbreaks of *Escherichia coli* O157:H7 infections among children associated with farm visits—Pennsylvania and Washington, 2000," *MMWR* (2001) 50:293–97.

16. "More than 1,000 sickened in deadly *E. coli* outbreak," *Orlando Sentinel* Sept. 18, 1999:A16.

17. CDC, "Listeriosis: technical information," www.cdc.gov/ncidod/dbmd/diseaseinfo/ listeriosis_t.htm, accessed Oct. 1, 2003.

18. Vugia et al. "Preliminary FoodNet data."

19. FDA CFSAN, *Bad Bug Book: Listeria monocytogenes*, www.cfsan.fda.gov/~mow/chap6. html 92); FDA CFSAN and USDA FSIS, "Preventing foodborne listeriosis," background document (1992), http://vm.cfsan.fda.gov/~mow/fsislist.html; and FDA CFSAN and

USDA FSIS, "*Listeria monocytogenes* risk assessment questions and answers," www. foodsafety.gov/~dms/lmr2qa.html.

20. P.A. Beloeil, P. Fravalo, C. Chauvin, et al., "*Listeria* spp. contamination in piggeries: comparison of three sites of environmental swabbing for detection and risk factor hypothesis," *J Vet Med B Infect Dis Vet Public Health* (2003) 50:155–60.

21. CDC, "Listeriosis: general information," www.cdc.gov/ncidod/dbmd/diseaseinfo/ listeriosis_g.htm, accessed Oct. 1, 2003. Outbreaks caused by vegetables are so rare that the Partnership for Food Safety Education does not even list vegetables as a source for *Listeria*. Partnership for Food Safety Education, "Organisms that can bug you: causes and symptoms" (2000), www.fightbac.org/content/view/14/21/.

22. B. Rowland, "Listeriosis," at Health A to Z: Your Family Health Site (2002), www. healthatoz.com/healthatoz/Atoz/ency/listeriosis.jsp.

23. USDA FSIS, "Listeriosis and pregnancy: what is your risk?," foodborne illness and disease fact sheet (2001), www.fsis.usda.gov/Fact_Sheets/Listeriosis_and_Pregnancy_ What_is_Your_Risk/index.asp; and Mayo Clinic, "Meningitis," www.mayoclinic.com/ health/meningitis/DS00118/dsection=3, accessed Dec. 28, 2005.

24. CDC, National Center for Infectious Diseases, "Fact sheet: variant Creutzfeldt-Jakob Disease" (2003), www.cdc.gov/ncidod/dvrd/vcjd/factsheet_nvcjd.htm, accessed Dec. 29, 2005.

25. National Cattlemen's Beef Association, "Cattlemen dispute report saying mad cow disease may be in U.S.," www.beefusa.org/newscattlemendisputereportsayingmadcowdisease maybeinus9864.aspx, accessed Dec. 29, 2005.

26. National Creutzfeldt-Jakob Disease Surveillance Unit, "CJD figures" (Edinburgh: Western General Hospital), www.cjd.ed.ac.uk/figures.htm, accessed May 2, 2006; and M. Enserink, "After the crisis: more questions about prions," *Science* (2005) Dec. 16:1756–58.

27. N. Hunter, "Scrapie and experimental BSE in sheep," *Br Med Bull* (2003) 66:171–83; and M.E. Bruce, "TSE strain variation," *Br Med Bull* (2003) 66:99–108.

28. D. Taylor, "Inactivation of the BSE agent," *C R Acad Sci III* (2002) 325:75–76; and H. Baron and S.B. Prusiner, "Prion diseases," in D.O. Fleming and D.L. Hunt, eds., *Biological Safety, Principles and Practices* (Washington, DC: ASM Press, 2000), pp. 187–208.

29. CDC, "BSE (bovine spongiform encephalopathy), or mad cow disease," www.cdc.gov/ ncidod/dvrd/bse/, accessed Nov. 22, 2005.

30. A. Binkley, "Canada, U.S. grapple with new BSE recommendations," *Food Chem News* (2003) 45:27; and J. Marsden, *AMI Fact Sheet: Meat Derived by Advanced Meat Recovery* (Washington, DC: American Meat Institute, 2002), www.amif.org/ FactSheetAdvancedMeatRecovery.pdf.

31. MedicineNet.com, "Variant Creutzfeldt-Jakob Disease (vCJD)," www.medicinenet.com/ variant_creutzfeldt-jakob_disease/article.htm, accessed Dec. 23, 2005.

32. USDA FSIS, "Beef."

33. CDC, "Salmonellosis: technical information," www.cdc.gov/ncidod/dbmd/diseaseinfo/ salmonellosis_t.htm, accessed Dec. 28, 2005; and CDC, "Salmonellosis"; and USDA ERS, "Briefing room: economics of foodborne disease—*Salmonella*" (2003), www.ers.usda. gov/Briefing/FoodborneDisease/Salmonella.htm, accessed Oct. 16, 2003.

34. University of Washington School of Medicine, "Reiter's syndrome," www.orthop. washington.edu/uw/tabID__3376/ItemID__52/mid__10313/Articles/Default.aspx, accessed Oct. 17, 2003.

35. Vugia et al., "Preliminary FoodNet data."

36. USDA FSIS, *Focus on Beef* (2002), www.fsis.usda.gov/oa/pubs/focusbeef.htm; K. Todar, "*Salmonella* and salmonellosis," in *Todar's Online Textbook of Bacteriology*

(University of Wisconsin–Madison, Department of Bacteriology), www.textbookofbacteriology.net/salmonella.html, accessed Dec. 29, 2005; J. Ackerman, "Food: How safe? How altered?," *Nat Geog* May 2002:2–50; and D. Cole, L. Todd, and S. Wing, "Concentrated swine feeding operations and public health: a review of occupational and community health effects," *Env Health Perspect* (2000) 108(8):685–89.

37. USDA, Animal Plant Health Inspection Service (USDA APHIS), *Info Sheet: Salmonella in United States Feedlots* (Fort Collins, CO, 2001), www.aphis.usda.gov/vs/ceah/ncahs/nahms/feedlot/feedlot99/FD99salmonella.pdf; S.J. Wells, P.J. Fedorka-Cray, D.A. Dargatz, et al., "Fecal shedding of *Salmonella* spp. by dairy cows on farm and at cull cow markets," *J Food Prot.* (2001) 64:3–11; J.S. Bailey, N.J. Stern, P. Fedorka-Cray, et al., "Sources and movement of *Salmonella* through integrated poultry operations: a multistate epidemiological investigation," *J Food Prot* (2001) 64(11):1690–97; and USDA APHIS, "Shedding of *Salmonella* by finisher hogs in the U.S." (1997), www.aphis.usda.gov/vs/ceah/ncahs/nahms/swine/swine95/sw95salm.pdf.

38. D.G. White, S. Zhao, R. Sudler, et al., "The isolation of antibiotic-resistant *Salmonella* from retail ground meats," *New Engl J Med* (2001) 345:1147–53.

39. CDC, "*Salmonella enteriditis*: general information" (2005), www.cdc.gov/ncidod/dbmd/diseaseinfo/salment_g.htm, accessed Dec. 29, 2005.

40. USDA National Animal Health Monitoring System, "*Salmonella enterica serotype enteritidis* in table egg layers in the U.S." (2000), www.aphis.usda.gov/vs/ceah/ncahs/nahms/poultry/layers99/lay99se.pdf, p. 1.

41. P.S. Holt, "Molting and *Salmonella enterica* serovar *enteritidis* infection: the problem and some solutions," *Poult Sci* (2003) 82:1008–10.

42. FDA CFSAN, *Bad Bug Book: Staphylococcus aureus*, www.cfsan.fda.gov/~mow/chap3.html; Predicala et al., "Bioaerosols in swine barns"; and M. Hajmeer, "*Staphylococcus aureus*" (Davis, CA: University of California, Department of Population Health and Reproduction, 2005), www.vetmed.ucdavis.edu/PHR/PHR150/2005/aureus.PDF.

43. FDA CFSAN, *Bad Bug Book: Staphylococcus aureus.*

44. Partnership for Food Safety Education, *Ten Least Wanted Foodborne Pathogens* (2003), www.fightbac.org/10least.cfm, accessed July 1, 2004.

45. CDC, *Toxoplasma Infection* (Division of Parasitic Diseases, 2003), www.cdc.gov/ncidod/dpd/parasites/toxoplasmosis/2004_PDF_Toxoplasmosis.pdf; and J.D. Kravetz and D.G. Federman, "Toxoplasmosis in pregnancy," *Am J Med* (2005) 118:212–16.

Photo Credits

We thank the following sources for their courtesy in providing images for this book.

- Animal Welfare Institute, www.awionline.org – *p. xii*
- Compassion Over Killing – *pp. 119 (top), 122*
- Corbis – *cover photo; pp. 117, 129*
- Courtesy of Cynthia Goldsmith, Jacqueline Katz, and Sherif R. Zaki, Centers for Disease Control and Prevention – *p. 67*
- Coronary Health Improvement Project – *p. 27*
- Dale Farm Limited – *p. 155 (top)*
- Augustine G. DiGiovanna, Salisbury University (© 2004, used with permission) – *p. 30*
- Farm Sanctuary – *pp. 113, 120, 121, 124, 125*
- Janet Green – *pp. 93, 111*
- Courtesy *Not Just For Vegetarians—Delicious Homestyle Cooking, The Meatless Way* by Geraldine Hartman – *pp. 22, 57, 149*
- Jason Hoverman, University of Pittsburgh – *p. 85*
- Barbara Hunt – *p. 147*
- Courtesy of the Kowalcyk family – *p. 63*
- Milk Processor Education Program – *p. 157*

- National Aeronautics and Space Administration – *p. 98*
- National Cancer Institute (Renee Comet, photographer) – *p. 18*
- National Dairy Council – *p. 44*
- National Institutes of Health, National Institute of Allergy and Infectious Disease, Rocky Mountain Laboratories – *p. 62*
- National Food Administration of Sweden – *p. 155 (bottom)*
- National Park Service – *p. 137*
- Photodisc – *frontispiece, p. 147*
- Poplar Spring Animal Sanctuary – *p. 166*
- Joe Skorupa, U.S. Fish and Wildlife Service – *p. 95*
- U.S. Congress, Architect of the Capitol – *p. 151*
- U.S. Department of Agriculture – *pp. 103, 133, 164*
- U.S. Department of Agriculture, Agricultural Research Service – *frontispiece* (Michael Macneil, photographer), *pp. vii, xiv, 9, 19, 31, 33, 35, 36, 37, 39, 42, 47, 50, 51, 53, 72, 73, 119 (bottom), 131, 138, 143, 153, 162 (top)*
- U.S. Department of Agriculture, Natural Resources Conservation Service – *pp. ix, xi, xiii, 10, 13, 65, 69, 71, 75, 76, 77, 79, 80, 83, 87, 91, 94, 101, 106, 112, 116, 139, 159, 161, 162 (bottom), 163*
- Prof. Kurt Wüthrich, ETH Zürich – *p. 66*

Index

Abbott Laboratories, 69
Acid rain, 108
Advanced meat recovery (AMR), 174
Advertising, 157
Agricultural Health Study (National Cancer Institute), 54
Agricultural practices. *See also* Fertilizers; Livestock production; Soil
affecting non-farm animals, 136–38
compaction and, 78
environmental damage from, xii, 77, 94–100
erosion and, 76–77
global, xiii
of small vs. large farms, 152
AgriProcessors Inc., 134
Air pollution
ammonia and, 104–06
effects of, 106–07
fertilizers and, 108–09, 161
from manure, 104–06, 109–10, 158
methane and, 107
nitric oxide and nitrogen dioxide and, 108

nitrous oxide and, 107–08
odor and, 110
overview of, 103–04
particulate matter and, 106, 109, 111
pesticides and, 112, 161
recommendations to prevent, 159, 161–62
volatile organic compounds and, 106, 110
Alatorre, José, 103
Algal blooms, 98
Alpha-linolenic acid, 11, 51
American Academy of Pediatrics, 57, 69
American Beekeeping Federation, 138
American Cancer Society, 42, 57, 144
American Diabetes Association, 144
American Dietetic Association, 147
American Grass Fed Beef, 10
American Heart Association, 11, 46, 57, 144
American Institute for Cancer Research, 144
American Meat Institute, 114
American Medical Association, 69
American Public Health Association, 69, 158